SpringerBriefs in Materials

For further volumes:
http://www.springer.com/series/10111

Waqar Ahmed · Htet Sein · Mark J. Jackson
Christopher Rego · David A. Phoenix
Abdelbary Elhissi · St. John Crean

Chemical Vapour Deposition of Diamond for Dental Tools and Burs

 Springer

Waqar Ahmed
Htet Sein
David A. Phoenix
Abdelbary Elhissi
St. John Crean
School of Medicine and Dentistry
Institute of Nanotechnology
    and Bioengineering
University of Central Lancashire
Preston
UK

Mark J. Jackson
Center for Advanced Manufacturing
Purdue University
West Lafayette, IN
USA

Christopher Rego
School of Science & the Environment
Manchester Metropolitan University
Manchester
UK

ISSN 2192-1091   ISSN 2192-1105   (electronic)
ISBN 978-3-319-00647-5   ISBN 978-3-319-00648-2   (eBook)
DOI 10.1007/978-3-319-00648-2
Springer Cham Heidelberg New York Dordrecht London

Library of Congress Control Number: 2014941686

Printed on acid-free paper

Springer is part of Springer Science+Business Media (www.springer.com)

Preface

Research and development in the area of machining and cutting tool development is well established in the advanced manufacturing industries. The theories associated with machining traditional materials such as metals, ceramics, polymers, composites, have been developed over the last century to a fairly high level of sophistication. However, the same cannot be said for the development of tools for the medical industry.

The purpose of this Springer 'brief' is to show the embryonic stages of development of highly engineered tools for the medical sector that can equally be applied to all areas of medicine and not just to the development of tools for dentistry. The brief also discusses the merits of applying well-established principles of physical understanding to the problems associated with cutting of biological materials, and is composed of eight chapters that focus on tools, teeth and their environment, the advantage of using coated tools, the synthesis, properties and application of diamond thin films, challenges associated with depositing thin diamond films to flat and curved surfaces, controlling diamond morphology and structure, and the assessment of tool performance using established measures of machinability.

The brief should excite all manner of interested professionals including dentists, orthodontists, surgeons, general practitioners, engineers, physicists, general scientists and students who are actively engaged in studying medicine and biomedical engineering. The authors wish to stimulate the readers' curiosity in order to enable them to conduct their own research, advance their particular frontier in medicine, and innovate and commercialize current and future knowledge generated in this area of medical engineering by experts in the field.

May 2014

Contents

Abstract

Chemical vapour deposition (CVD) is a powerful technology for depositing thin films of diamond onto complex substrates such as dental tools and dental burs. The advantages of diamond for these applications are that it is the hardest material known to mankind, has a high thermal conductivity and a high wear resistance. When a diamond coating is applied to dental tools and burs the performance and life are considerably enhanced. This brief book outlines the CVD technologies used, the diamond growth mechanism, film properties and assesses their impact on the performance and life of diamond coated tools compared to uncoated tools and conventional burs.

Abstract

Chapter 1
Diamond Synthesis, Properties and Applications

Abstract Diamond is an ideal material for numerous applications such as cutting tools such as dental burs and drills due to its unique combination of chemical, mechanical and thermal properties. The most widely used method of growth diamond is chemical vapour deposition (CVD) namely hot filament and microwave plasma processes. The use of vertical filament chemical vapour deposition (VFCVD) process has been developed to uniformly coat complex shaped tools and is described in detail. The growth characteristics and film properties are described for use on dental burs and drills.

Keywords Diamond · Thin films · Chemical vapour deposition · CVD · Growth mechanism · Film properties

1.1 Diamond

For most people diamond is a coveted and famously desired gemstone used for jewellery due to its shiny lustre and attractiveness (May 1995). The word diamond comes from the Greek language and has the meaning "invincible". Its applications are vast arising from its excellent and unique combination of physical, chemical, electrical and optical properties. For example, for cutting tools its extreme hardness, high thermal conductivity and very low coefficient of friction are highly desirable (Angus 1991; Yoder 1994). Its chemical inertness and biocompatibility makes it suitable for biomedical implants such as artificial heart valves, hip and knee prosthesis (Yoder 1994; Field 1992).

Diamond is crystalline and is one of three carbon allotropes along with graphite and fullerenes e.g. (C_{60}). It is the most stable of all of the allotropes with the highest density. Natural diamond, obtained through mining is scare and very expensive, is of limited use hence scientists have been curious to find ways of making high quality synthetic diamond in the laboratory that could be used for a wide range of applications (Field 1992; Anusavice 2012; May 1995). For over a half a

W. Ahmed et al., *Chemical Vapour Deposition of Diamond for Dental Tools and Burs*, SpringerBriefs in Materials, DOI: 10.1007/978-3-319-00648-2_1, © The Author(s) 2014

century a high-pressure high-temperature (HPHT) process has been used. A US patent was granted in 1962 for work on synthetic diamond using chemical vapour deposition (CVD) (Eversole 1962). This powerful technique produces polycrystalline, microcrystalline or nanocrystalline diamond as a thin film, which adheres to a suitable substrate, or as freestanding wafers (May 1995). Free standing may be used for optical windows in space applications because diamond is radiation resistant.

1.2 Diamond as an Allotrope of Carbon

Carbon occurs in different physical forms known as diamond, graphite, fullerene, carbon nanotubes and graphene with various differences in bonding and structure. Graphite and diamond are the most common forms of crystalline carbon whilst graphene has only been discovered recently arousing great curiosity and with new applications being researched for this new wonder material (Geim and Novoselov 2007). The structure and bonding between the carbon atoms determine the properties of these allotropes. For example, graphite is soft and the layers can slide over one another. In contrast, diamond is the hardest natural material. Graphene occurs as a single sheet and potentially numerous applications including solar cells paving the way for an environmentally friendly and sustainable source of energy.

Graphite is thermodynamically stable structure for carbon at standard pressure and temperature (STP), which contains planes of threefold coordinated carbon atoms, bound by strong sp^2 bonds (Trucano and Chen 1975). In this allotrope the s orbital mixes with two p orbitals forming three new sp^2 hybridised orbital making a planar triangle. The four outer shell electrons are arranged such that three electrons occupy the sp^2 orbitals forming three sigma bonds and one electron stays in the p_z orbital a π bond. The electrons in the π bond are responsible for its good electrical conductivity. A flat sheet of connected carbon atoms is formed. Covalent bonds are present within the layers of graphite and weak van der Waals forces between the layers, each separated by 3.40 Å (Mao et al. 2003). Graphite consists of sheets of hexagonal rings with weak interactions between the sheets. Hence, the layers of carbon rings can slide over each other making graphite a good dry lubricant. This makes it useful for many applications such as bearings in space components.

In diamond, the C–C bonds are of the same length, with equal bond angles at 109.47°. It consists of repeating structural units made of 8 atoms arranged in a cube. Diamond, however, has each carbon bonded to four other carbons in a tetrahedral arrangement (Aleksenskii et al. 1999). Diamond has different properties. It can be cleaved along its planes. However, it cannot flake apart into layers due to the tetrahedral structure of carbon atoms. Diamond has sp^3 configuration of atoms of carbon at 25 °C and 1 atmosphere.

The two most common allotropes diamond and graphite differs in energy by 0.02 eV per atom. In addition, there is a difference in ΔH_f^θ values by ~2.9

kJ mol^{-1} (Herman 1952). However, diamond and graphite has a large activation energy barrier at room temperature and pressure. This stops the diamond from converting to graphite and vice versa. However, under aggressive conditions such as at high temperature and ion bombardment the transformation between diamond and graphite occurs at a faster rate.

In 1985 Robert F. Curl, Harold W. Kroto, and Richard Smalley discovered new forms of carbon called fullerenes in which the atoms are arranged in closed shells for which they won the Nobel Prize in Chemistry in 1996 (Kroto et al. 1991). Closed shells with various number of carbon atoms have been found in numerous carbon structures. The most abundant and stable being clusters with 60 atoms known as C_{60} which is a polyhedron consisting of 20 hexagons and 12 pentagons. The structure has been named as "buckminsterfullerene" or the "buckyball" (Kroto 1988).

Recently, there has been considerable research work on new forms of carbon nanostructures called carbon nanotubes (CNTs). These have been observed in the 1950s as filaments of carbon. Iijima in 1991 found them to be hollow tubes of carbon with a diameter of the order of one nanometer and these tubes have been found to be several microns in length (Iijima and Ichihashi 1993). They exist in several forms including single walled CNTs (SWCNT), double walled CNTs (DWCNT) and multi walled CNTs (MWCNT). SWCNT is a layer of graphite one atom thick called graphene. A sheet of graphene can be wrapped into a seamless cylinder with either open or closed ends. DWCNTs consist of two sheets of graphene rolled to form two concentric cylinders. MWCNTs consist of multiple concentric layers of graphene forming tubes of several concentric cylinders. These have numerous potential new applications as novel material (Iijima 2002; Bianco et al. 2005).

In 2010 Geim and Novoselov were awarded the Nobel Prize in physics for their pioneering ground breaking work on graphene, which is a single atomic layer of sp^2 bonded carbon a two-dimensional atomic crystal with a range of unusual properties. It exhibits extraordinary crystal and electronic properties revealing a cornucopia of new physics and numerous potential applications (Geim 2011; Hancock 2011). For example, graphene thermal conductivity and mechanical stiffness may rival the remarkable in-plane values for graphite (~3,000 W m^{-1} K^{-1} and 1,060 GPa, respectively). The fracture toughness could be comparable to carbon nanotubes and with remarkable electronic properties. These properties may be harnessed by homogeneously incorporating graphene into other materials such as polymers to form composite materials and structures (Stankovich et al. 2006). Chemical modification of graphene for use in composite materials could extend the range of its properties and applications. For example, by using chemically modified graphene, polystyrene-graphene composites exhibit percolation threshold of around ~0.1 volume per cents for electrical conductivity at 25 °C which is much lower than the lowest value reported for any carbon based composites except for carbon nanotubes. At 1 % by volume these composites exhibit a conductivity of ~0.1 S m^{-1}, suitable for range of electrical applications. This approach enables a broad new class of graphene-based materials to be used in numerous applications.

Table 1.1 Physical
properties of diamond

Property	Value
Hardness (kg mm^{-2})	10,000
Strength, tensile (GPa)	>1.2
Strength, compressive (GPa)	>110
Density (g cm^{-3})	3.52
Young's modulus (GPa)	1.22
Thermal expansion coefficient (K^{-1})	0.0000011
Thermal conductivity (W cm^{-1}K^{-1})	20.0

Table 1.2 Some applications
of diamond related to its
properties

Properties of diamond	Applications
Hardest known materials	Cutting tools
Chemical inertness	Electrochemical sensors
High thermal conductivity	Heat spreaders
Biological inertness	In vitro applications (coatings/ sensors)
High resistivity (insulator)	Electronic devices, sensors
Semiconducting when doped	Electronic devices, sensors
Negative electron affinity	"Cold cathode" electron sources

1.3 Properties and Applications of Diamond

Diamond has been deposited on various substrates including silicon, which is a semiconductor and metals such as copper and insulators such as SiO_2. They range from single crystals, polycrystalline to amorphous materials. When diamond thin films are grown with CVD methods new technological applications emerge because a variety of substrates can be coated. CVD operates at relatively high temperature compare to PVD. The difference in temperature yields either polycrystalline diamond films in CVD and the softer diamond-like carbon films in the case of PVD. This expands the potential application areas of diamond when compared to natural or HPHT-synthetic diamond. It is anticipated that this technology will have a major economic impact on sectors such automotive, consumer, defence and space applications (Spear and Dismukes 1994).

Diamond, as mentioned earlier, has an extreme combination of properties. These include good electrical insulation, high thermal conductivity, and a low dielectric constant that makes it highly suitable for packaging electronic devices and technologies used in multichip manufacture. Diamond's high hardness value and wide optical band gap renders it suitable for optical applications. Combined with chemical inertness and high wear resistance diamond thin films are suitable protective coatings in cutting tools and metal working industries. Table 1.1 summarises the physical properties of diamond (Field 1992). The electrical, structural, and optical properties of diamond are at the most extreme end of the spectrum. These properties render diamond to be potentially the most useful material across many fields of science and technology. Table 1.2 shows the variety of potential applications of diamond thin films as a function of its extreme properties (Field 1992).

1.4 Synthesis of Diamond

High Pressure High Temperature (HPHT) process and Chemical Vapour Deposition (CVD) are two main methods for synthesising diamond in crystalline and thin film forms respectively. The CVD processes enable the production of diamond coatings of excellent quality on flat or 3-D substrates and also synthesis of freestanding shapes of diamond. Free-standing diamond can be obtained by etching away the substrate using various chemical etching procedures (Anthony and Fleischer 1993; Frey and Simpson 1994; Elmazria et al. 2003). Diamond needs to be formed under conditions that are thermodynamically favourable. From the carbon phase diagram it is evident that heating carbon under pressure can form diamond. Hence, the high-pressure high-temperature technique was developed to produce 'industrial' diamond (Liang et al. 2005). A cell operating at high pressures up to pressures of tens of thousands of atmospheres is used to compress the graphite (G) at >2,000 K with a metal catalyst. This is left until diamond crystals form in the high-pressure cell.

1.4.1 Historical Perspective of Diamond Synthesis

From historical, technological and scientific perspective diamond is fascinating natural material (Angus et al. 2002). Sir Isaac Newton was the first to characterise diamond and determine it to be of organic origin whilst in 1772, the French chemist established that the product of the complete combustion of diamond was limited to carbon dioxide (Hartley 1947). English chemist (Tennant 1797) showed that diamond combustion products were no different than those of coal or graphite and resulted in 'bound air' namely CO_2 (Weeks 1933). Later, the advent of the x-ray diffraction technique enabled scientists Sir William Henry Bragg and his son Sir William Lawrence Bragg to determine the crystalline nature of the carbon allotropes, which were found to be cubic (diamond) and hexagonal (graphite) (Bragg and Bragg 1913). Since diamond is the densest carbon phase, it is feasible that high pressure producing a smaller volume and therefore a higher density, may convert other forms of carbon into diamond. During the 19th and 20th centuries chemical thermodynamics became better understood making it feasible to explore diamond stability in various pressure-temperature regimes (Berman and Simon 1955).

In 1955, these studies led to the development of a high pressure-high temperature (HPHT) process (DeVries 1987). Three problems arose in making HPHT diamond in the laboratory. First, it is difficult to achieve the extreme pressure that is required. Secondly, even when such a high pressure has been achieved, a very high temperature is required to convert graphite to diamond at a reasonable rate. Finally, if diamond is obtained, the grains are very small. In order to achieve large single crystal diamond yet another set of experimental conditions are needed. During that period however, less well known is the development of low-pressure diamond growth. This phase depends on choosing suitable experimental conditions (Butler et al. 1993). The diamond surface converts to graphite spontaneously.

Further nucleation and growth of graphitic species also occurs (Butler and Windischmann 1998).

The most successful process for low-pressure growth of diamond has been chemical vapour deposition from energetically activated hydrocarbon/hydrogen gas mixtures. CVD is a process whereby a thin solid film is synthesised from the gaseous phase via a chemical reaction. The most ancient example of a material deposited by CVD is pyrolytic carbon. Blocher pointed out that incomplete oxidation during the burning of firewood resulted in soot (Blocher 1966). A very early patent was granted for the preparation of carbon black using CVD (Donnet 1993; Bachmann et al. 1991; Bachmann and van Enckevort 1992). Later a patent was also granted for the next major application of CVD, which involved coating fragile filaments made from carbon (Blocher 1966). Several years after this, metal deposition processes for improving lamp filaments were presented (Spear 1989). Prior to late 1930s several techniques were described for refractory metals preparation. Silicon tetrachloride ($SiCl_4$) and hydrogen (H_2) mixture was used to deposit silicon films for microelectronic application (Joffreau et al. 1988).

$$SiCl_4(g) + 2H_2(g) \rightarrow Si(s) + 4HCl(g) \qquad (1.1)$$

This led to the development of silicon-based photocells in 1946 as well as rectifiers (Photo-Cell 1946). The electronics industry has driven many of the developments of CVD processes. Consequently, a huge amount of literature on CVD is available. It was the chemical vapour deposition from carbon-containing gases, which enabled Eversole to grow diamond at low pressures (Eversole 1958). In the initial experiments, carbon monoxide was used as a source gas to deposit diamond on a diamond seed crystal. However, in subsequent experiments, methane and other carbon-containing gases were used as well as cyclic growth/etch procedures to grow diamond yet removing co-deposited graphite. In all of his studies, it was necessary to use diamond seeds in order to initiate diamond growth. The deposits were identified as diamond by density measurements, chemical analysis and diffraction techniques. Diamond was also synthesised successfully at high pressure in 1954. However, the important difference was that Eversole grew diamond on pre-existing diamond nuclei (Derjaguin et al. 1975). In 1956 diamond whiskers were grown using a metal-catalysed vapour-liquid-solid process (Spitsyn et al. 1981). A little later, epitaxial growth was the subject of considerable interest. In addition, theoretical studies on the nucleation of diamond and graphite were carried out. Angus and co-workers worked primarily on diamond CVD on diamond seed crystals from hydrocarbons hydrogen mixtures (Angus 1991). The *p*-type diamond was grown from methane/diborane gas mixtures and they studied the rates of diamond and graphite grown from the methane/hydrogen gas mixtures and ethylene. It was suggested that atomic hydrogen increased the rate of graphite etching compared to diamond etch rates. Low energy electron diffraction (LEED) studies (Messier et al. 1987) were done on a (Anusavice)-diamond surface. The hydrogen atoms are believed to be responsible for terminating the dangling bonds normal to the growing surface. Without hydrogen complex surface structures are formed. At high temperatures above 1200 K the carbon atoms move about rapidly on the surface. Other work (Angus and Hayman 1988; Spear 1989) suggested when hydrogen is present diamond growth is enhanced (Kamo et al. 1983). Conversely, hydrogen addition to the hydrocarbon precursor suppresses the graphite formation relative to diamond. The resulting diamond films are of a higher quality. However, graphitic carbons can also nucleate on the surface to suppress diamond growth. Therefore graphitic deposits need to be removed using atomic hydrogen or oxygen; this sequence may be repeated. The importance of hydrogen is now widely accepted. Diamond growth rates of the order of 0.1 μm h^{-1} had been achieved by the mid 70s. These growth-rates were too low for commercial applications.

1.4.2 Modern Era of Diamond Growth

In 1974 Japanese researchers made the first disclosures of a method for rapid diamond growth at low pressures. Further work in 1982, the same researchers reported the synthesis of diamond at much higher rates using hot filament CVD (HFCVD), microwave CVD (MWCVD) and direct current (DC) discharges at growth rates of several μm/h (Matsumoto et al. 1982a, b). In the absence of diamond seed crystals these processes produced growth of individual facets. Deryagin et al. (1976) had reported high rate diamond growth but did not give the detailed process parameters employed in the study. The deposition techniques mentioned are based on the generation of atomic hydrogen above the growth surface. Although the chemical vapour deposition of diamond from hydrogen rich/hydrocarbon-containing gases has been the most successful method of diamond synthesis, numerous other methods have been attempted with varying degrees of success with ion beam methods being the most successful (Liou et al. 1990). In 1971 hard carbon films were first deposited using a carbon ion beam. This process produced diamond like carbon (DLC) films. Spencer et al. (2008) used an ion beam of carbon in 1976 to form finely divided polycrystalline diamond. They employed very low ion energies between 50 and 100 eV, and this was followed by diamond growth work using ion implantation by Freeman et al. (1978). To take advantage of the unique characteristics of diamond as a material for the construction of solid-state devices, basic scientific understanding of the various experimental process techniques and in particular the introduction and activation of dopants must be obtained. The potential of diamond as a material for solid-state devices has been reviewed extensively (Davis 1993).

1.5 Development of CVD Technology

Various CVD techniques have been developed to produce hard and wear resistant coatings of materials such as diamond, DLC, boron, borides, cBN, carbides and nitrides have become important cutting tool (Bhushan et al. 1993; Sein et al. 2004). A CVD system is composed of a gas delivery system, a reactor (microwave or thermal), pumping system and an exhaust (Ahmed et al. 2000). The CVD reactor is a critical component and provides several functions including:

- It must transport the reactant and diluent gases to the substrate,
- Provide sufficient activation energy to the reactants,
- Maintain specific system pressure and temperature profiles,
- Allow film deposition to proceed,
- Remove the gaseous by-products.

1.5.1 Types of Diamond CVD Processes

Several CVD systems, processes are used to deposit thin films for a range of applications (Jones and Hitchman 2009). There are several different approaches to activating the gas phase precursor gases for CVD of diamond and these include the following:

1. Plasma-Enhanced CVD

Plasma is a complex mixture containing ions, radical and neutral species occurring in excited state. Plasmas generated by various forms of electrical discharges or induction heating have been employed for diamond growth. The plasma generates atomic hydrogen and the necessary carbon precursors for diamond growth. The efficiencies of the different plasma processes vary from method to method. Three plasma frequency regimes will be discussed. These are *Microwave* plasma CVD which typically uses excitation frequencies of 2.45 GHz, *Radio-frequency* (RF) plasma excitation, which employs frequencies of usually 13.56 MHz, and *Direct-current* (DC) plasmas which can be run at low electric powers (a "cold" plasma) or at high electric powers which create an *arc* or a *thermal* plasma.

2. Microwave-Plasma Enhanced CVD

MWCVD has been used more extensively, and more successfully than any other plasma-based method for the growth of diamond films. Microwave plasmas are different from other plasmas in that the microwave frequency can oscillate electrons. Collision of electrons with gaseous atoms and molecules generates a high degree of ionisation. This method of diamond film growth has a number of advantages over the other plasma methods of diamond film growth. Microwave deposition, is a cleaner process because it does not have electrodes and therefore avoids contamination of the films due to electrode erosion. Furthermore, the microwave discharge at 2.45 GHz, being a higher frequency process than the RF, produces a higher plasma density with higher energy electrons. This effort results in higher concentrations of atomic hydrogen and other hydrocarbon radicals resulting in efficient high quality diamond growth. In addition, as the plasma is confined to the centre of the deposition chamber as a ball, so carbon deposition onto the walls of the chamber is prevented.

Table 1.3 Deposition parameters used in the growth of diamond films by HFCVD

Gas mixture	Total pressure (Torr)	Temperature (K)	
		Substrate	Filament
CH_4 (0.5–2.0 %)/H_2	10–30	1,000–1,400	2,200–2,500

3.   RF Plasma Enhanced CVD

Generally, *rf* power is applied to two electrodes to create plasma in an inductively coupled or a capacitively coupled parallel plate arrangement. The growth of diamond crystals and thin films using inductively coupled RF plasma methods and capacitively coupled methods have been reported extensively in the literature. A high power in the discharge leading to greater electron densities is necessary for efficient diamond growth. However, the use of higher power results in physical and chemical sputtering from the reactor walls leading to contamination of the diamond films. The advantage of RF plasmas is that they can easily be generated over much larger areas than microwave plasmas but the method is not routinely used for the deposition of diamond films.

4.   DC Plasma Enhanced CVD

In this method, plasma in a H_2-hydrocarbon mixture is excited by applying a DC bias across two parallel plates, one of which is the substrate. DC plasma enhanced CVD has the advantage of being able to coat large areas as the diamond deposition area is limited by the electrodes and the DC power supply. In addition, the technique has the potential for very high growth-rates. However, diamond films produced by DC plasmas were reported to be under high stress and to contain high concentrations of hydrogen as well as impurities resulting from plasma erosion of the electrodes.

5.   Hot Filament CVD (HFCVD)

Atomic hydrogen can be produced by the passage of H_2 over a metal filament, made of tungsten, molybdenum, rhenium and tantalum, heated to temperatures between 2,000–2,500 K. When atomic hydrogen was added to the hydrocarbon typically with a C/H ratio of ~0.01, it was observed that diamond could be deposited while graphite formation was suppressed. The generation of atomic hydrogen during diamond HFCVD enabled the following:

- A dramatic increase in the diamond deposition rate to approximately 1 $\mu m\ h^{-1}$
- The nucleation and growth of diamond on non-diamond substrates.

HFCVD is simple and cheap and has become widely used in industry. Table 1.3 outlines typical deposition parameters used in the growth of diamond films by this technique. A wide variety of refractory materials have been used as filaments. These include tungsten, tantalum, molybdenum and rhenium due to their high melting point and high electron emissivity.

1.5.2 CVD Mechanisms

In the HFCVD process, typically the precursors are entering reactor in premixed composition (CH_4:H_2) and transported by carrier gas. The actual decomposition of reactant occurs within activated hot zone from where the products are transported towards the surface. The surface processes and reactions will determine the resulting film properties and deposition performances. The filament temperature and choice of process pressure strongly influence the degree of precursor fragmentation (radical formation i.e. CH_3 and H), gas phase nucleation and deposited film morphology. The generic process can be seen as two-step, transport and deposition stages, as shown below.

Various mechanisms for diamond growth (Van Enckevort et al. 1993; Banholzer 1992; Tsang et al. 1997; Banholzer 1992; Lee et al. 1999) have been postulated. Thermodynamic near-equilibrium is established in the gas-phase at the filament surface. At temperatures around 2,300 K, molecular hydrogen dissociates into atomic hydrogen and the H atoms are able to convert methane into methyl radicals, which are generally regarded as the active diamond precursor species, and thereafter into acetylenic species and other hydrocarbons, which are stable at these elevated temperatures. Atomic hydrogen and the neutral and radical hydrocarbon species then diffuse to the substrate surface. Although the gaseous species generated at the filament are in equilibrium at the filament temperature, the species are at a super-equilibrium concentration when they arrive at the much cooler substrate. The reactions, which generate these high temperature species close to the filament where there is a relatively high concentration of hydrogen atoms, proceed faster than any reactions, which decompose these species during the transit time from the filament to the substrate. Consider the equilibrium between methane and acetylene:

$$CH_4 + H \leftrightarrows CH_3 + H_2 \leftrightarrows {}^1/_2 C_2H_2 + 2H_2 \qquad (1.2)$$

At the filament surface, the reaction is immediately driven to the right, creating methyl radicals and thereafter acetylene. Subsequently, the CH_3 radicals and others hydrocarbon species diffuse to the substrate. Thermodynamic equilibrium at the substrate temperature of ~1,100 K calls for the formation of methane, but the reverse reaction proceeds much slower. Solid carbon precipitates on the substrate to reduce the super-equilibrium concentration of species such as CH_3 in the gas-phase. The allotrope diamond is "stabilised" relative to graphite by a concurrent super-equilibrium concentration of atomic hydrogen. This simple explanation emphasises the importance of reaction kinetics in diamond synthesis by HFCVD. The mechanism of diamond CVD is summarised in Fig. 1.1. Not only do the H atoms play a crucial role in diamond CVD by driving the formation of active hydrocarbon species but they also prevent both the reconstruction of the growing diamond surface and the formation of graphitic nuclei.

To deposit high quality diamond and at acceptable growth rates the C:H ratio in precursor gas mixture is crucial. Increasing the proportion of carbon results in higher concentrations of CH_3 at the growing diamond surface and hence increased growth rates. However, higher concentrations of carbon also lead to poorer quality films, as there is insufficient H to etch away non-diamond carbon deposits. If the concentration

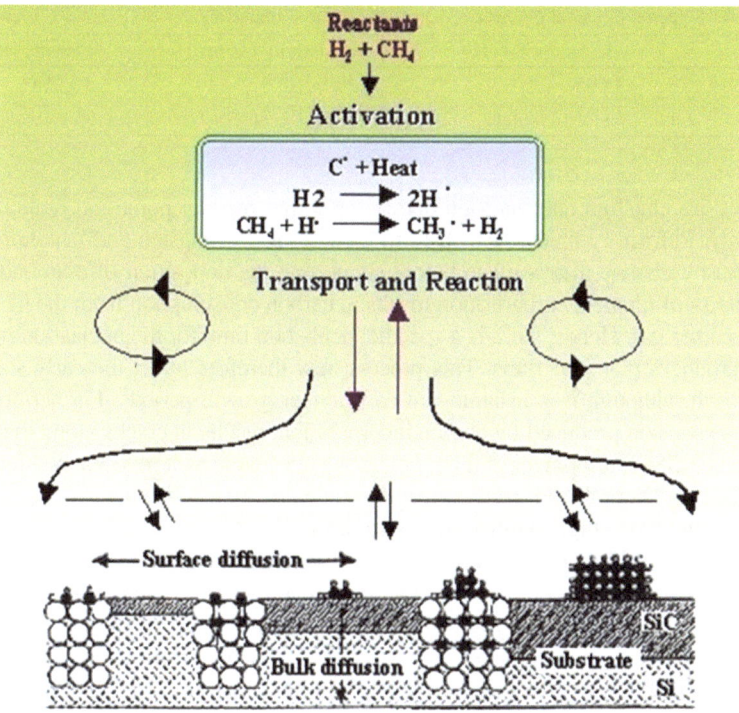

Fig. 1.1 Schematic of the CVD diamond growth mechanism (Afzal et al. 1998)

of H at the growing surface is too high then diamond will be etched resulting in very low growth rates or no diamond growth at all. In HFCVD, the transport of the active species and their mixing with the surrounding gas takes place. The concentration and thermal gradients are the major driving forces. To grow uniform film coatings the temperature of the substrate and the local species concentration need to be consistent. At the surface various processes occur. These include adsorption, surface diffusion, nucleation, island formation and continuous film formation with the adjacent islands merging. Undesirable species are desorbed and transported away from the surface into the main gas stream. The atomic hydrogen also plays a key role in etching away the surface; however it etches graphite at a faster rate than diamond resulting in a greater proportion of diamond in the films compared to graphitic species. An important feature of HFCVD is that the growth step is separate from the film activation step, which ultimately allows independent control of both growth and activation temperatures.

1.5.3 Filament Characteristics

It is obvious that in a process such as HFCVD the filament plays a critical role (Sommer and Smith 1990). The key features of the filament characteristics and assembly have been summarised (Ledermann et al. 2001). The most commonly used filament

Table 1.4 Selected physical data for prospective filament materials

Material	Melting point (K)	Resistivity (Ω m)	Density (kg m^{-3})
W	3,695	39×10^{-8}	19,254
Ta	3,293	63×10^{-8}	16,670

materials are tantalum and tungsten due to their high melting point and high electron emissivity. Refractory metals, which form carbides (e.g., tungsten and tantalum) typically, must carburise their surface before supporting the deposition of diamond films. The process of filament carburisation results in carbon consumption from the hydrocarbon precursor gas. Hence, there is a specific incubation time for the nucleation process, which produces diamond films. This process may therefore affect the early stages of film growth, although it is insignificant over longer growth periods. Furthermore, the volume expansion produced by carbon incorporation, results in cracks along the length of the wire. The development of these cracks is undesirable as it reduces the lifetime of the filament but does not adversely affect the quality of the resulting films.

In addition to a high melting point suitable filament materials must also have an appreciable electron emissivity to cause dissociation of molecular hydrogen and ultimately initiate the growth process. For these reasons tungsten (W, melting point 3,695 K) and tantalum (Ta, melting point 3,293 K) filaments are typically used in HFCVD processes. The physical data for tungsten and tantalum is shown in Table 1.4. In this research work Ta filaments were used because there is less contamination of the diamond films than when W filaments are used. The differences in the chemical natures and physical properties of the filament materials affect precursor activation and decomposition hence diamond deposition may vary slightly given otherwise constant deposition parameters.

A wide range of literature on HFCVD based diamond deposition places the low filament temperature limit at 2,200 K and the upper limit at 2,700 K for successful diamond growth. The filament carburisation process proceeds according to the following steps:

1. Transport of CH_4 from the gas-phase to the metal surface and subsequent physical adsorption;
2. CH_4 decomposition on the metal surface where the resulting carbon and hydrogen atoms are chemisorbed;
3. Evolution of gaseous H, H_2 and CH_X;
4. Transformation of the adsorbed C atoms into the dissolved state;
5. Diffusion of carbon atoms into the metal lattice to form sub carbides (M_2C) and carbides (MC).

If the filament temperature falls below a critical value, depending on the carbon concentration, a carbon film forms on the filament. This film deactivates the filament surface (i.e. the filament temperature decreases) and the production of atomic hydrogen is rapidly reduced. If the filament power is not increased immediately to dissolve or vapourize the carbon film, it will grow thicker which leads to a further decrease in the filament temperature. Emissivity measurements, aim at measuring the spectral emissivity of the filaments. The filament activity is highest if the filament surface is totally free of carbon. The changes in filament emissivity, resistance and power consumption have been attributed to the deposition and etching of carbon occurring at the surface.

1.5.4 Substrates for Growing Diamond

Several criteria dictate the choice of substrates (May 1995a, b; May et al. 1995). They include the following:

- The melting point of the substrate must be higher than the diamond deposition temperature (>750 °C).
- The substrate should have similar thermal expansion coefficient to diamond to prevent film cracking during growth.
- The substrate material is able to form a carbide layer onto which diamond can adhere and deposit.

There are three main groups of substrates based their reaction rates with carbon (Lux and Haubner 1996; Haubner and Lux 1996) and these are described below.

1. *Substrates with no carbon solubility*

The group of metals such as copper, tin, lead, silver and gold as well as non-metals, such as germanium, sapphire, diamond itself, and graphite, although in the latter case etching will occur concurrently with diamond growth. These substrates are not suitable for diamond growth because they cannot form a carbide layer onto which diamond can nucleate and subsequently growth; however these may be used to grow free standing diamond films since they can be readily removed the substrate after deposition.

2. *Substrates with high carbon solubility*

Here, the substrate acts as a carbon sink, whereby deposited carbon dissolves into the metal surface to form a solid form. This indicates that large amounts of carbon to be transported into the bulk, leading to a temporary decrease in the surface C concentration, delaying the beginning of nucleation. Those metals include Pt, Pd, Rh, Fe, Ni, and Ti. In this type of substrates the material absorbs carbon like a sink. The carbon dissolves and forms a solid solution. Large quantities of carbon enter into the substrate causing diamond to nucleate. The growth of diamond starts after carbon saturates the substrate. The physical properties of the substrate are modified significantly and produce diamond films with poor adhesion since the carbon species diffuses rapidly into the surface making it difficult to form a continuous film.

3. *Substrates with limited carbide formation*

These include metals such as Ti, Zr, Hf, V, Nb, Ta, Cr, Mo, W, Fe, Co, Ni, Y, Al and certain other rare earth metals. In some metals, such as Ti, the carbide layer continues to grow during diamond deposition and can become hundreds of μm thick. Some thick interfacial carbide layers may severely affect the mechanical properties, and hence the utility of CVD diamond coatings on these materials. Non-metals, such as B or Si, and Si-containing compounds such as SiO_2, quartz, and Si_3N_4 also form carbide layers. Substrates composed of carbide themselves, such as SiC, WC, and TiC are also particularly amenable to diamond deposition; materials in which carbon has limited solubility. In cases such as Ti, the carbide layer continues to grow to hundreds of μm during diamond deposition, which can severely affect the mechanical properties thus limiting applications of diamond. However, some common non-metals, such as Si, B, SiO_2, and Si_3N_4, form limited carbide layers onto which diamond can nucleate and deposit. Carbide substrates including SiC, WC and TiC are common for diamond deposition for a number of applications.

Once of the most widely used substrates for diamond growth is silicon. It has the following characteristics:

- High melting point (1,683 K)
- Forms a localised carbide layer
- Low thermal expansion coefficient.

Tungsten and molybdenum display similar properties to silicon for diamond growth and are very popular substrate materials.

1.6 Diamond Nucleation Process

For obvious reasons the initial stage of diamond nucleation on a non-diamond substrate is critical (Lifshitz et al. 2004; Lee et al. 2000; Liu and Dandy 1996). It affects the uniformity and morphology of the diamond films deposited by CVD. Abrading and seeding with diamond powder or immersing in diamond paste containing small crystallites processed in an ultrasonic bath is commonly used in diamond film deposition technique. The method promotes nucleation of diamond crystals onto the substrate surface by creating a high density of nucleation sites, which reduces the induction time. Growth of diamond begins when individual carbon atoms nucleate onto the surface to initiate the beginning of a sp^3 tetrahedral lattice. The gas activation is done either by Hot Filaments, Microwave or Radio-Frequency-Plasmas. The most crucial parameter in all these processes, besides a carbon source, is the presence of large amounts of atomic hydrogen. The role of atomic hydrogen in the process is:

- Creation of active growth sites on the surface
- Creation of reactive growth species in the gas-phase
- Etching of non-diamond carbon (like graphite) graphitic, sp^2, precursors will be explored.

Nucleation of diamond is a critical and necessary step in the growth of diamond thin films, because it strongly influences diamond growth, film quality and morphology. Growth of diamond begins when individual carbon atoms nucleate onto the surface to initiate the beginnings of sp^3 tetrahedral lattice. There are two different types of diamond growth in which homo-epitaxial growth and hetero-epitaxial growth are most commonly used.

1.6.1 Homo-Epitaxial Growth

Diamond growth is achieved by a variety of processes using very different means of gas activation either HFCVD or microwave method. These processes show how process variables (such as pressure, gas activation, substrate/filament temperature, characteristic diffusion length, gas composition and flow rates) influence diamond growth rates and diamond quality. When using natural diamond substrates, there is a ready-made template for the required tetrahedral structures, and the diamond lattice is just extended atom-by atom whereas dangling diamond can be bonded as deposition process proceeds.

1.6.2 Hetero-Epitaxial Growth

This process uses non-diamond substrates where there is no template for the C atoms to follow, and those C atoms that deposit in non-diamond forms are immediately etched back to the gas phase by reaction with atomic hydrogen. There are a few problems to deal before the initial induction period of diamond deposition; the substrate must undergo pre-treatment prior to deposition in order to reduce the induction time for nucleation and to increase the density of nucleation sites. In this study only hetero-epitaxial growth of diamond deposition is applied. Two methods commonly used to apply this pre-treatment:

The first method is manual abrading or seeding or with diamond powder or immersing in diamond paste containing small crystallites processed in an ultrasonic bath to enhance nucleation. It is essential that a nucleation mechanism for diamond is available on non-diamond substrates in which pre-abrasion reduces the induction time for nucleation by increasing the density of nucleation sites. The abrasion process can be achieved by mechanically polishing the substrate with abrasive grit, usually diamond power of 0.1–10 μm-particle size.

The second method is bias-enhanced nucleation (BEN) in which a negative or positive bias is applied to the filament relative to the substrate. In this study negative bias-enhanced nucleation (BEN) is used for hetero-epitaxial growth of diamond on non-diamond surface. BEN is used as a nucleation step in this work and information on the experimental process and mechanism are given in Chap. 6.

1.7 Conclusions

Owing to the gas phase nature of CVD the deposition of diamond film should be uniform even on complex substrates if the temperature and species concentrations are consistent. Therefore, variable and complex shaped surfaces such as micro drills and dental burs can be coated evenly with diamond films. CVD process is carried out at relatively high temperatures and therefore there are limitations in the types of the substrates that may be coated.

Diamond films have been deposited mainly on selected small areas of drill bits, cutting tools and bearing surfaces. Even though the technology has developed at an accelerate rate deposition onto complex shapes is still challenging. Therefore reactors need to be designed appropriately to give uniform temperature and concentration gradients across the sample surfaces to be coated. This work aims to coat thin diamond films on substrates that are challenging because they have complex 3-D geometries, the substrate materials require surface pre-treatment to grow high quality diamond films.

References

Afzal A et al (1998) HFCVD diamond grown with added nitrogen: film characterization and gas-phase composition studies. Diam Relat Mater 7(7):1033–1038

Ahmed W et al (2000) CVD diamond: controlling structure and morphology. Vacuum 56(3):153–158

Aleksenskii AE et al (1999) The structure of diamond nanoclusters. Phys Solid State 41(4):668–671

Angus JC (1991) Diamond and diamond-like phases. Diam Relat Mater 1(1):61–62

Angus JC, Hayman CC (1998) Low pressure, metastable growth of diamond and "diamond like" phases Science 241(4868):913–921

Angus JC et al (2002) A short history of diamond synthesis. Diam Films Handbook 1–17

Anthony TR, Fleischer JF (1993) Substantially transparent free standing diamond films. US Patent No 5,278,931

Anusavice KJ (2012) Standardizing failure, success, and survival decisions in clinical studies of ceramic and metal–ceramic fixed dental prostheses. Dent Mater 28(1):102–111

Bachmann PK et al (1991) Towards a general concept of diamond chemical vapour deposition. Diam Relat Mater 1(1):1–12

Bachmann PK, van Enckevort W (1992) Diamond deposition technologies. Diam Relat Mater 1(10):1021–1034

Banholzer W (1992) Understanding the mechanism of CVD diamond. Surf Coat Technol 53(1):1–12

Berman R, Simon SF (1955) On the graphite-diamond equilibrium. Zeitschrift für Elektrochemie, Berichte der Bunsengesellschaft für physikalische Chemie 59(5):333–338

Bhushan B et al (1993) Tribological properties of polished diamond films. J Appl Phys 74(6):4174–4180

Bianco A et al (2005) Applications of carbon nanotubes in drug delivery. Curr Opin Chem Biol 9(6):674–679

Blocher JM (1966) II Crushing strength at individual pyc-coated particles as a function of silicon content, Google Patents

Bragg WH, Bragg WL (1913) The structure of the diamond. Nature 91:557

Butler JE, Windischmann H (1998) Developments in CVD-diamond synthesis during the past decade. MRS Bull 23(09):22–27

Butler JE et al (1993) Thin film diamond growth mechanisms [and comment]. Philos Trans R Soc Lond Ser A: Phys Eng Sci 342(1664):209–224

Davis RF (1993) Diamond films and coatings Noyes Publications(USA):435

Derjaguin BV et al (1975) Structure of autoepitaxial diamond films. J Cryst Growth 31:44–48

Deryagin BV et al (1976) Diamond crystal synthesis on nondiamond substrates. Sov Phys Dokl

DeVries RC (1987) Synthesis of diamond under metastable conditions. Annu Rev Mater Sci 17(1):161–187

Donnet JB (1993) Carbon black: science and technology, CRC Press

Elmazria O et al (2003) High velocity SAW using aluminum nitride film on unpolished nucleation side of free-standing CVD diamond. Ultrason, Ferroelectr Freq Control, IEEE Trans 50(6):710–715

Eversole WG (1958) Liquid-gas contacting apparatus. US Patent No 2,819,887

Eversole WG (1962) Synthesis of diamond. US Patent No 3,030,188

Field JE (1992) The properties of natural and synthetic diamond. Academic Press, London

Freeman JH et al (1978) Epitaxial synthesis of diamond by carbon-ion deposition at low energy. Nature 275:634–635

Frey RM, Simpson M (1994) Method for making free-standing diamond film. US Patent No 5,314,652

Geim AK (2011) Nobel lecture: random walk to graphene. Rev Mod Phys 83(3):851

Geim AK, Novoselov KS (2007) The rise of graphene. Nat Mater 6(3):183–191

Hancock Y (2011) The 2010 nobel prize in physics—ground-breaking experiments on graphene. J Phys D Appl Phys 44(47):473001

Hartley H (1947) Antoine Laurent Lavoisier 26 August 1743–8 May 1794. In: Proceedings of the Royal Society of London. Series B, Biological Sciences, pp 348–377

Haubner R, Lux B (1996) On the formation of diamond coatings on WC/Co hard metal tools. Int J Refract Metal Hard Mater 14(1):111–118

Herman F (1952) Electronic structure of the diamond crystal. Phys Rev 88(5):1210

Iijima S (2002) Carbon nanotubes: past, present, and future. Physica B 323(1):1–5

Iijima S, Ichihashi T (1993) Single-shell carbon nanotubes of 1-nm diameter

Joffreau PO et al (1988) Low-pressure diamond growth on refractory metals. Int J Refract Hard Met 7(4):186–194

Jones AC, Hitchman ML (2009) Chemical vapour deposition: precursors, processes and applications. R Soc Chem 490–530

Kamo M et al (1983) Diamond synthesis from gas phase in microwave plasma. J Cryst Growth 62(3):642–644

Kroto H (1988) Space, stars, C60, and soot. Science 242(4882):1139–1145

Kroto HW et al (1991) C60: buckminsterfullerene. Chem Rev 91(6):1213–1235

Ledermann A et al (2001) Influence of gas supply and filament geometry on the large-area deposition of amorphous silicon by hot-wire CVD. Thin Solid Films 395(1):61–65

Lee S et al (1999) CVD diamond films: nucleation and growth. Mater Sci Eng: R: R 25(4):123–154

Lee ST et al (2000) A nucleation site and mechanism leading to epitaxial growth of diamond films. Science 287(5450):104–106

Liang ZZ et al (2005) Synthesis of HPHT diamond containing high concentrations of nitrogen impurities using NaN_3 as dopant in metal-carbon system. Diam Relat Mater 14(11):1932–1935

Lifshitz Y et al (2004) Visualization of diamond nucleation and growth from energetic species. Phys Rev Lett 93(5):056101

Liou Y et al (1990) The effect of oxygen in diamond deposition by microwave plasma enhanced chemical vapor deposition. J Mater Res 5(11):2305–2312

Liu H, Dandy DS (1996) Diamond chemical vapor deposition: Nucleation and Early Growth Stages, Elsevier

Lux B, Haubner R (1996) Diamond deposition on cutting tools. Ceram Int 22(4):347–351

Mao WL et al (2003) Bonding changes in compressed superhard graphite. Science 302(5644):425–427

Matsumoto S et al (1982a) Vapor deposition of diamond particles from methane. Jpn J Appl Phys 21(4A):L183

Matsumoto S et al (1982b) Growth of diamond particles from methane-hydrogen gas. J Mater Sci 17(11):3106–3112

May P (1995a) Synthetic diamond: Emerging CVD science and technology: Edited by Harl E. Spear and John P. Dismuhes, pp. 663. Wiley, Chichester, 1994. ISBN 0 4715 3589 3. Endeavour 19(1):48

May PW (1995b) CVD diamond: a new technology for the future? Endeavour 19(3):101–106

May PW et al (1995) CVD diamond-coated fibres. Diam Relat Mater 4(5–6):794–797

Messier R et al (1987) From diamond-like carbon to diamond coatings. Thin Solid Films 153(1):1–9

Photo-Cell ANES (1946) Applied physics. J Appl Phys 17:215

Sein H et al (2004) Performance and characterisation of CVD diamond coated, sintered diamond and WC–Co cutting tools for dental and micromachining applications. Thin Solid Films 447:455–461

Sommer M, Smith FW (1990) Activity of tungsten and rhenium filaments in CH4/H2 and C2H2/H2 mixtures: importance for diamond CVD. J Mater Res 5(11):2433–2440

Spear KE (1989) Diamond—ceramic coating of the future. J Am Ceram Soc 72(2):171–191

Spear KE, Dismukes JP (1994). Synthetic diamond: emerging CVD science and technology. Wiley, New York

Spencer EG et al (2008) Ion-beam-deposited polycrystalline diamondlike films. Appl Phys Lett 29(2):118–120

Spitsyn BV et al. (1981) Vapor growth of diamond on diamond and other surfaces. J Cryst Growth 52 Part 1(0):219–226

Stankovich S et al (2006) Graphene-based composite materials. Nature 442(7100):282–286

Tennant S (1797) On the nature of the diamond. By Smithson Tennant, Esq. FRS. Philosophical Transactions of the Royal Society of London, pp 123–127

Trucano P, Chen R (1975) Structure of graphite by neutron diffraction. Nature 258(5531):136–137

Tsang RS et al (1997) Examination of the effects of nitrogen on the CVD diamond growth mechanism using in situ molecular beam mass spectrometry. Diam Relat Mater 6(2):247–254

Van Enckevort WJP et al (1993) CVD diamond growth mechanisms as identified by surface topography. Diam Relat Mater 2(5):997–1003

Weeks ME (1933) The discovery of the elements chronology. J Chem Educ 10(4):223

Yoder MN (1994) Synthetic diamond: emerging CVD Science and technology. Wiley, New York

Chapter 2
Dental Tools, Human Tooth and Environment

Abstract The construction of dental burs and hand pieces are described in this chapter of the book. Initially, the chapter describes the characteristics of the bur focusing on design and the effect of different rake angles on cutting, then the chapter focuses on the design and construction of elements of the hand piece itself. The finals aspect explains the direction of research that needs to be taken in order to develop a hand piece that has been optimized for use with different dental materials.

Keywords Dental burs · Dental hand piece · Air turbine · Mechanical design · Fluid dynamics · Human tooth

2.1 Introduction

Dental burs are used for cutting hard tissues such as bone or tooth (O'Brien 1997). They are usually made of stainless steel, diamond grit or particles and tungsten carbide. Burs are rotary drill bits of various shapes and dimensions used by the dentist in his dental drill incorporating an air turbine. The development of the dental bur dates back approximately 300 years. The drill bit for dental work has revolutionised the field of dentistry.

A dental bur consists of three parts, which include bur head, bur neck and shank. The bur head contains the blades that rotate at high speed to cut and remove material from the teeth (Sein et al. 2002). They can be positioned at various angles to remove unwanted material such as plaque from difficult to get at areas. The positive rake angle, which is also called the acute angle, has a sharper blade, which dulls rapidly during use. The obtuse angle produce negative rake angle, which makes the bur stronger and last longer. Dental technicians also use burs to prepare dental materials in the laboratory.

In order to appreciate the role dental burs are playing in dentistry it is informative here to give some background into the oral environment. Normally an adult

W. Ahmed et al., *Chemical Vapour Deposition of Diamond for Dental Tools and Burs*, 19
SpringerBriefs in Materials, DOI: 10.1007/978-3-319-00648-2_2, © The Author(s) 2014

human has 32 teeth, which are aligned in two dental arches. First one is called maxillary arches, and the second is called madibular arches. A single tooth consists of a crown with neck, one or more cusps, and a root (Seely et al. 2007). The clinical crown is the part of the tooth exposed in the oral cavity, and is entirely enamel. The pulp cavity is filled with nerves, blood vessels, and connecting tissue. Root canal is situated in the pulp cavity within the root. The nerves and blood vessels of the tooth have inlet and outlet routes to the pulp through a hole at the point of each root. It is commonly called epical foramen. The pulp cavity is surrounded by living, cellular, calcified tissue called dentine. The dentine of the crown is covered by non-living, acellular substance called enamel, which is very hard and protects the tooth against abrasion and acid produced from bacteria in the mouth. When the tooth becomes weaker and soft tissues break down and can cause pain. Untreated cavities cause diseases such as dental caries and abscesses requiring treatment by a qualified dentist (Seely et al. 2007).

The modern day dental bur has the ability to allow dentists to work more efficiently and accurately with less pain. The future goal of dentistry is to maintain and improve the dental health of patients. This can be achieved in a number of ways such as disease prevention, improved mastication efficiency, and relieves pain, speech and appearance. These objectives can be achieved by employing better dental tools that are durable, long lasting, biocompatible, and cost effective.

A major aim of this research project is to evaluate the performance of CVD diamond coated dental burs and micro drill cutting tools within simulated clinical and laboratory conditions. The goal for these surface engineered cutting tools is to improve the product quality, cutting performance and life expectancy (Sein et al. 2002; Ahmed et al. 2004).

2.2 Dental Burs

A dental bur or dentist's drill is used to remove decayed tooth material before applying a filling. Dental burs may come in various shapes and sizes designed for specific applications and can rotate up to 500,000 rpm. Steel is a widely used material for making burs, coated with a hard WC coating. They can also be entirely made by sintered WC (Trava-Airoldi et al. 1996a, b; Carvalho et al. 2007). In recent years the bur has been diamond coated using CVD (Trava-Airoldi et al. 1996a, b). There are numerous shapes of burs manufactured for varying applications, cutting and drilling abilities.

2.2.1 Dental Bur Construction and Geometry

Each dental bur comprises of three main parts; shank, shaft and bur head (the cutting edge). Since the bur is composed of a single parent material (with the

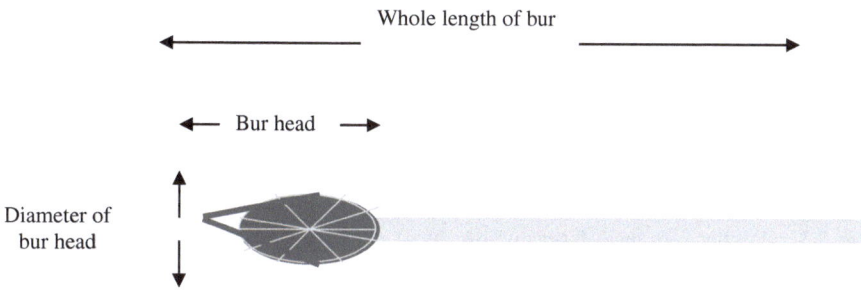

Fig. 2.1   Schematic of the various parts of a dental bur

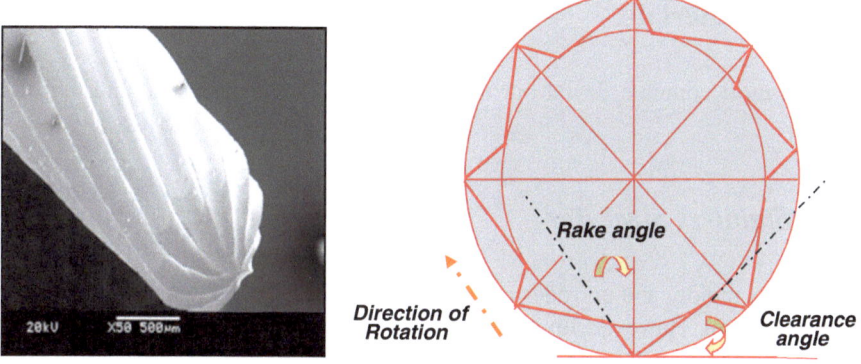

Fig. 2.2   Geometry of a typical tungsten carbide bur and schematic diagram showing the rake angle relative to the cutting direction (Sein et al. 2004a, b)

possibility of a coating along the cutting surfaces) the choice of this material is important (see Fig. 2.1). It must provide a good cutting edge, but also be able to withstand the forces endured by the shaft throughout the cutting process.

The basic design of an eight bladed fissure bur is indicated in Fig. 2.2. Most burs have a negative rake angle. Those with a positive rake angle are designed mainly for cutting soft materials (e.g. acrylics). This is designed to remove material during cutting to prevent the tool from clogging with cut material or chips.

When the rake angle of a bur blade is too steep, subsurface damage occurs in the tooth. These weak areas become sites for later bacterial infection. When the rake angle is decreased, a more gentle action is achieved and less residual subsurface damage occurs to the tooth but life of the bur is decreased due to the acute angle of the cutting tips. The rake angles come in two different varieties positive and negative (Fig. 2.3).

If the leading edge of the blade is behind the perpendicular the angle is by definition positive. Examples of positive angle instruments are finecut files, hedstrom files and most dental burs when run normally.

If leading the edge of blade is ahead of perpendicular, the angle is by definition, negative. Examples of negative angle instruments are reamers and diamond burs

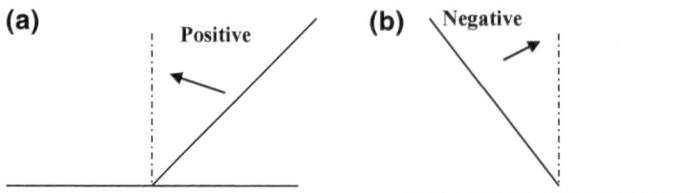

Fig. 2.3 **a** Positive rake angle, **b** negative rake angle

and burnishing burs when runs backward. The geometrical design features also impact on the performance and wear rates (Polini et al. 2004).

2.3 Bur Materials

The two most common dental bur materials used are stainless steal and tungsten carbide.

2.3.1 Stainless Steel Bur

Steel wears rapidly and corrodes easily. This does not make it appear to be a good choice for burs. However, limiting the use of steel burs to lower speeds significantly reduces the wear rate associated with the cutting process. The main corrosive environment experienced by a bur is during the sterilisation process, although the conditions in the oral cavity will also contribute (Dietz et al. 2000). Stainless steel provides a less efficient cutting edge than carbon steel, but is more corrosion resistant. Stainless steel is also a cheap material, which makes it suitable for burs provided that they are considered single-use and classed as disposable burs.

2.3.2 Tungsten Carbide WC-Co Bur

Tungsten carbide is an extremely hard material, which makes it suitable for use at very high speeds. Unfortunately, associated with the additional hardness of WC, relative to stainless steel, is an increase in brittleness. Therefore only the blades of a bur should be made of tungsten carbide (the shank being made of steel). Sintering the carbide blades onto the steel shank joins the two components. Tungsten carbide is also suitable for burs designated for use at lower speeds, when they are to be used many times. Instruments made from tungsten carbide are much more expensive than their steel equivalents, but compensate for this by their increased working life (Sein et al. 2003). WC burs contains cobalt as a binder, which provides additional toughness to the tool. Most WC-Co dental burs contain about 6 % Co as a binder material. There are two parameters of importance these

Fig. 2.4 SEM of a conventional sintered diamond bur showing lateral and *top views* (Sein et al. 2004a, b)

are the Co/WC ratio and the WC particle size. These parameters have a significant impact on the control of the bulk material properties (Jian et al. 2004). In general, course grained WC combined with a high % Co give better shock resistance and impact strength. However, finer grained WC and thus larger surface area and lower % Co gives harder and greater wear resistance (Jian et al. 2004).

Therefore to achieve optimum performance premature failure needs to be avoided whilst at the same time achieving a higher wears resistance (Yared et al. 2001).

2.3.3 Sintered Polycrystalline Diamond Dental Bur

As discussed in the previous section the diamond bur is a favoured instrument of both dentists and dental technicians (Sein et al. 2004a, b, 2006). Since the 1950s conventional diamond bur technology uses small diamond particles bonded onto the substrate using a binder matrix material (Fig. 2.4); its use is however limited due to variation in the grain sizes and grain shapes, problems in the automation of the fabrication process and the short tool life. Sterilisation of the instruments also decreases the cutting effectiveness and results in diamond particle loss causing oral contamination (Siegel and von Fraunhofer 1999).

2.4 Operation Conditions

Many different designs of dentals trepan are available in the marketplace. They must however have common features (see Fig. 2.5). These include the following:

- Drill bit
- Motor
- Couplings
- Hand piece

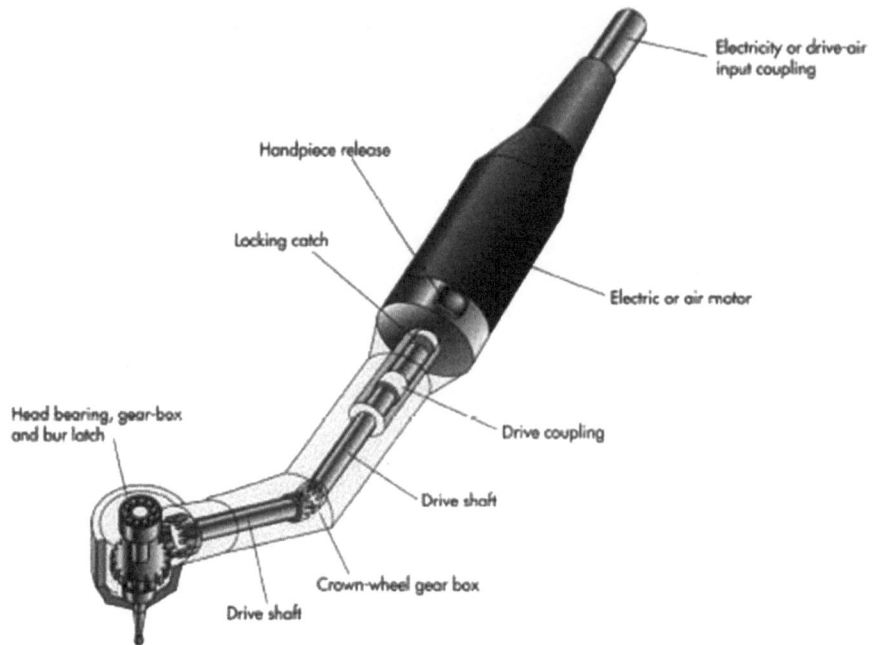

Fig. 2.5 High speed dental handpiece. http://www.gs-dental.com/UpFile/2009313114114.jpg

An air turbine activates the high speed drilling up to 0.5 million rpm. Many operations require slower speeds and therefore dental drills also have secondary motors. For example, polishing, finishing, and soft tissue drilling require the slower rotational speeds. Electric motors and air-driven motors are widely used.

The hand piece is a well-designed, slender and lightweight tube-shaped device. This connects the drill bit and a motor that provides the driving force during its use. During the early phases of their developments, dental drill components were somewhat delicate. However, health risks and legislation has forced changes in the way in which the dentil drill is designed. The hand piece can now withstand high-pressure steam sterilization. The drill bit or bur is a highly important component and needs to be robust, durable and be able to rotate at very high speeds. These high speeds generate a lot of heat, which needs to be removed using water-cooling. Modern devices have addition features such as illumination devices to aid the dentist in performing cutting, drilling and grinding operations.

Each bur is designed to operate at an optimum speed and intended for use in a specific hand piece motor combination. The optimum speed is also dependent upon the nature of the material being cut. Cutting speed can be separated into two categories each using different burs and designed for different types of operations.

2.4.1 Low Speed Hand Piece

A miniature compressed air or electric motor is used to drive low speed hand pieces. They operated with rotational speeds of up to 4,000 rpm. The burs used in low speed operations are used to trim dentures and remove decayed tissue. These burs may be made of stainless steel.

2.4.2 High-Speed Hand Piece

Air turbine hand piece can run up to 500,000 rpm, although they are rarely used at such high speeds. General operating speeds are approximately 20,000–50,000 rpm (varying with the diameter of the bur). The burs used at high speeds are mainly diamond-coated or made of tungsten carbide. High speeds are used in the removal of enamel, dentine and old fillings. When operating at such high speeds, a built in water spray is used as a coolant, protecting both the bur and tooth. The efficiency of micro cutting tools that are used in high-speed air turbine spindles depends on the rotational speed of the rotor. A high-pressure variation on the surface of the rotor causes the rotor to retard and this severely limits the reliability and durability of high-speed spindles. A variety of spindle designs are proposed and numerical simulations carried out for each design using CFD FLUENT^{TM} simulation software. The results revealed that changes in the rotor, inlet, and outlet geometries affect the pressure distribution on the rotor significantly. The optimum design was identified based on the lowest pressure variation on the rotor surface obtained from the FLUENT^{TM} results. Spinning the rotor at very high speeds provides a new direction in the development of dental cutting tools.

The major components of a high-speed air turbine spindle are: bearings, rotor, stator, and spindle shaft. To drive a high-speed spindle a motor, or compressor, is integrated with the spindle shaft. Bearings provide stability at high speeds to prevent chatter and poor surface finish and to permit accurate cutting tool paths. The speed of the spindle depends on the rotational speed of the rotor. The spindle shaft, rotor, and bearings must be held in the housing. High-pressure compressed air enters into the housing of the spindle from the compressor through a pneumatic connector. The compressed air enters the housing through the shaft and rotates the rotor of the spindle. The micro cutting tool, which is attached to the centre of the rotor, rotates with the speed of the rotor and cuts the workpiece more quickly than conventional spindles. The rotor is supported by an air bearing, which provides stability to the rotor and also transmits the necessary torque.

In high speed machining with high-speed spindles, the pressure variation on the rotor surface is of vital importance. The pressure coefficient is defined as the difference between the highest and lowest pressure on the rotor surface, normalized by the imposed inlet pressure. Pressure coefficient determines pressure variation exerted by high-speed compressed air on the rotor and for the optimum design of the rotor, the pressure coefficient should be as low as possible as the large values of pressure coefficient indicate high pressure variations, which could cause the severe imbalance of the load and deformation of the rotor and this generates failure of the rotor. Various designs of rotor for different rotational speeds are proposed and fluid analysis of rotors has been carried out with a computational fluid dynamics (CFD) software package called FLUENT^{TM}. Pressure coefficients of rotors were calculated and compared for different designs of rotor to determine the optimum design of the rotor.

2.4.2.1 Analysis of Fluid Flow

1. Assumptions
   The following assumptions were considered for the numerical solution of high-speed spindles (HSS) using computational fluid dynamics (CFD):

- HSS rotor rotational speed depends on the pressure of the compressed air entering from the compressor. The numerical simulation was carried out by a considering de-coupled system, i.e., for a given inlet pressure (60 psi), the rotating speed of the rotor was assumed as a constant value such as half-million rpm, i.e., the current study deals only with the fluid problem, not the fluid/structure problem;
- Steady-state simulation was assumed for all numerical simulations.

2. CFD Geometry Model
   The bearing component of the HSS was omitted in the CFD model. The outer diameter of the rotor is 0.3 inches (7.6 mm), inner diameter of the rotor is 0.092 inches (2.34 mm), the height along the z-direction is 0.1445 inches (3.6 mm) and the angle between the rotor blades is 90°. A cylindrical housing with a diameter of 0.31 inches (7.8 mm) was modelled around the rotor with a height of 0.1735 inches (4.4 mm) so that the rotor could rotate freely inside the housing. The spindle is driven by compressed air. Three inlets, with a diameter of 0.055 inches (1.4 mm), that make an angle of 120° with each other, were created around the housing. The inlets were created at an angle of 45°. An outlet for the air was created at the centre of the housing.

3. Fluid Model
   The air was considered as an ideal gas. The flow in HSS is turbulent. The turbulence is described by k-ε turbulent model, in which k is the turbulence kinetic energy and ε is the turbulence eddy dissipation. The total energy heat transfer model was considered as kinetic energy effects are important in the model.

Applied Boundary Conditions
The following boundary conditions were applied to the model:
   - Static pressure of 60 psi at the inlets;
   - Static pressure of 0 (zero) at the outlet;
   - A no-slip (moving) wall boundary condition on the rotor. A constant angular speed of the rotor was specified;
   - A no-slip (stationary) wall boundary condition on the housing and inlet surfaces.

Governing Equations

The governing equations of three-dimensional fluid flow were represented as:
Continuity Equation:

$$\frac{\partial(\rho U_i)}{\partial X_i} = 0 \tag{2.1}$$

Momentum Equation:

$$\frac{\partial\left(\rho U_i U_j\right)}{\partial x_j} = -\frac{\partial P}{\partial x_i} + \frac{\partial}{\partial x_j}\left(\mu_{eff}\frac{\partial U_i}{\partial x_j} + \mu_{eff}\frac{\partial U_j}{\partial x_i}\right) \tag{2.2}$$

where repeated indices imply summation from 1 to 3, ρ is density, U_i are the carte-sian velocity components, P is pressure, X_i are the coordinate axes, and μ_{eff} is the effective viscosity, which is defined as:

$$\mu_{eff} = \mu + \mu_t; \quad \mu_t = C_\mu \rho \frac{k^2}{\varepsilon} \tag{2.3}$$

where μ_t is the eddy viscosity, C_μ is a constant and is equal to 0.09, k is the tur-bulence kinetic energy and ε is the turbulence eddy dissipation. The turbulence model is given by:

$$\frac{\partial \rho k}{\partial t} + \frac{\partial}{\partial x_j}(\rho U_j k) - \frac{\partial}{\partial x_j}\left(\frac{\mu_{eff}}{\sigma_k}\frac{\partial k}{\partial x_j}\right) = \mu_t\frac{\partial U_i}{\partial x_j}\left(\frac{\partial U_i}{\partial x_j} + \frac{\partial U_j}{\partial x_i}\right)$$
$$- \frac{2}{3}\frac{\partial U_j}{\partial x_j}\left(\mu_t\frac{\partial U_j}{\partial x_j} + \rho k\right) - \rho\varepsilon \tag{2.4}$$

and

$$\frac{\partial \rho\varepsilon}{\partial t} + \frac{\partial}{\partial x_j}(\rho U_j\varepsilon) - \frac{\partial}{\partial x_j}\left(\frac{\mu_{eff}}{\sigma_\varepsilon}\frac{\partial \varepsilon}{\partial x_j}\right) = \frac{\varepsilon}{k}\left(C_{\varepsilon1}\left(\mu_t\frac{\partial U_i}{\partial x_j}\left(\frac{\partial U_i}{\partial x_j} + \frac{\partial U_j}{\partial x_i}\right)\right.\right.$$
$$\left.\left. - \frac{2}{3}\frac{\partial U_j}{\partial x_j}\left(\mu_t\frac{\partial U_j}{\partial x_j} + \rho k\right)\right) - C_{\varepsilon2}\rho\varepsilon\right) \tag{2.5}$$

where σ_k and σ_ε are k-ε turbulence model constants and are equal to 1.0 and 1.3 respectively, and $C_{\varepsilon1}$ and $C_{\varepsilon2}$ are equal to 1.45 and 1.92, respectively:
   Energy Equation:

$$\frac{\partial\left(\rho U_j h_{tot}\right)}{\partial X_j} = \frac{\partial\left(\lambda\frac{\partial T}{\partial X_j}\right)}{\partial X_j} + S_E \tag{2.6}$$

where h_{tot} is defined as the specific total enthalpy, which for the general case of variable properties is given in terms of the specific static enthalpy, h, by:

$$h_{tot} = h + \frac{1}{2}U^2; \quad h = h(p, T) \tag{2.7}$$

and S_E is the source term, which represents the work done by the viscous and pressure forces. The equation of state for an ideal gas is given as:

$$p = \rho RT \qquad (2.8)$$

where T is the temperature of the fluid and R is the gas constant.

Pressure Coefficient

The parameter "pressure coefficient" was defined to determine the pressure variation on the rotor,

$$\text{Pressure coefficient} = \frac{P_{max} - P_{min}}{P_{inlet}} \qquad (2.9)$$

where P_{max} = Maximum pressure exerted by air on the rotor, P_{min} = Minimum pressure exerted by air on the rotor, P_{inlet} = Air pressure at the inlet. Maximum pressure and minimum pressure exerted by air can be obtained by FLUENT^{TM}. Inlet pressure was used to non-dimensionalize the value of the pressure coefficient.

2.4.2.2 Experimental Results and Discussions

Flow topology and pressure variation of rotor geometries such as rotor with 90° blade angle, rotor with inlets, inclined at 45° to the z-axis of the rotor, and two-stage rotor are described and pressure coefficient values for all geometries are calculated. The optimum design of the high-speed spindle is identified based on the magnitude of the pressure coefficient.

Numerical Results

Numerical simulations of the spindle geometries were carried out using FLUENT^{TM}. Air was the ideal gas with an inlet pressure of 60 psi and an outlet pressure of zero, and the rotor was considered to be at no-slip conditions at the wall with specific rotational speeds such as half-million rpm, one million rpm, etc. The total number of control volumes for the numerical grid was approximately between 437,000 and 532,000 based on the geometry of the rotor, and the number of iterations for numerical simulations was 200. The maximum continuity loops are advantageous for achieving convergence especially for high-speed flows and the value of maximum continuity loops was specified as 2 for numerical simulations of all geometries under consideration. The governing equations such as continuity equation, momentum equation, energy equation, and equation of state were solved by FLUENT^{TM} to provide the pressure distribution on the rotor. The pressure values were obtained using FLUENT^{TM} and the pressure coefficient values were calculated using Eq. (2.9).

Initially, numerical simulations of rotor geometries such as the basic geometry of the rotor, rotor with fillets, rotor with 70° blade angle, and the rotor with

Table 2.1 Pressure coefficient values for rotor geometries such as basic geometry of the rotor, rotor with fillets, rotor with 70° blade angle and rotor with 90° blade angle for different rotational speeds

Rotational speed (RPM)	Basic geometry of rotor	Rotor with fillets	Rotor with 70° blade angle	Rotor with 90° blade angle
Half-million	0.11	0.10	0.09	0.07
One million	0.18	0.20	0.16	0.11
Two million	0.45	0.46	0.40	0.42
Five-Million	1.28	1.18	1.15	1.19

90° blade angle were carried out for different rotational speeds of the rotor such as half-million rpm, one million rpm, two million rpm, and five million rpm. The number of control volumes considered for the numerical simulation of basic geometry of the rotor was 437,273; for the numerical simulation of rotor with fillets it was 439,092; for the numerical simulation of rotor with 70° blade angle it was 488,786 and for the numerical simulation of rotor with 90° blade angle it was 520,324. The residual target for the convergence criterion was specified as 1e-6 and the maximum number of iterations considered for convergence control was 200 for the described geometries. The values of the non-dimensional pressure coefficients obtained form the numerical simulations are shown in Table 2.1 and displayed in Fig. 2.6.

From Table 2.1 it is observed that for a particular geometry as the rotating speed of the rotor increases, the pressure coefficient value increases because of the variation of pressure values on the surface of the rotor. Figure 2.7 shows the variation of pressure coefficient for the specified geometries with the rotational speeds of the rotor. From Table 2.1 it can be said that the pressure coefficient values of rotor with 90° blade angle are less than those of the other geometries of the rotor for the rotational speeds of half-million, one-million, and two-million rpm of the rotor which is in the typical range of micromachining operations. For two million and five million rpm rotational speed of rotor, the rotor with a 70° blade angle possessed pressure coefficient values that were less when compared to other geometries of the rotor. The geometry of the rotor with 90° blade angle was considered to be the optimum design compared to other geometries of the rotor for the rotational speeds that are in the typical range of micromachining conditions. Modifications in the spindle geometry were conducted by changing the number of blades of the rotor, changing the angle of the inlets, and by changing the number of inlets and outlets for the fluid. Numerical simulations of the rotor with twelve blades, rotor with three inlets inclined at an angle of 30°, rotor with three inlets inclined at an angle of 45°, and rotor with three inlets and three outlets were carried out. The number of control volumes associated with the numerical simulation of rotor with twelve blades it was 467,311; with that of rotor with three inlets inclined at an angle of 30° it was 513,116; with that of rotor with three inlets inclined at an angle of 45° it was 532,447 and with that of rotor with three inlets and three outlets it was 509,288. The residual target for satisfying the convergence criterion was

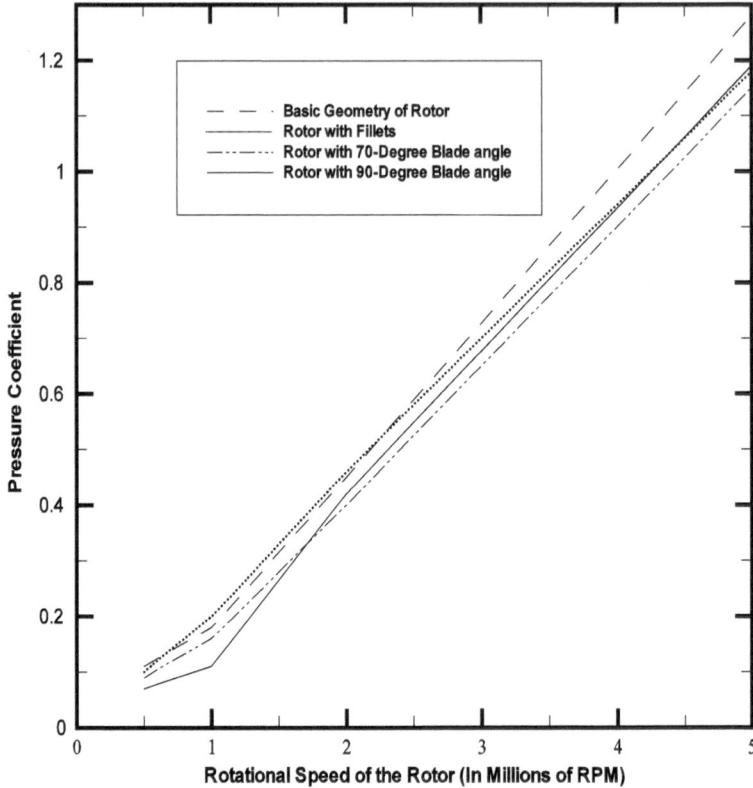

Fig. 2.6 Variation of pressure coefficient with rotating speed of the rotor for different geometries of the rotor such as basic geometry of the rotor, rotor with fillets, rotor with 70° blade angle and rotor with 90° blade angle. Reproduced with permission. Copyright retained by Inderscience Publishers

Table 2.2 Pressure coefficient values for rotor geometries of rotor with twelve blades, rotor with three inlets inclined at an angle of 30°, rotor with three inlets inclined at an angle of 45° and rotor with three inlets and three outlets

Rotational speed (RPM)	Rotor with twelve blades	Rotor with three inclined inlets at 30°	Rotor with three inclined inlets at 45°	Rotor with three inlets and three outlets
Half-million	0.09	0.10	0.08	0.36
One million	0.15	0.15	0.10	0.48
Two million	0.39	0.41	0.40	0.44
Five million	1.07	1.14	1.12	1.15

Reproduced with permission. Copyright retained by Inderscience Publishers

specified as 1e-6 and the maximum number of iterations specified for convergence control was 200. The pressure coefficient values obtained for these geometries for different rotational speeds of the rotor are shown in Table 2.2.

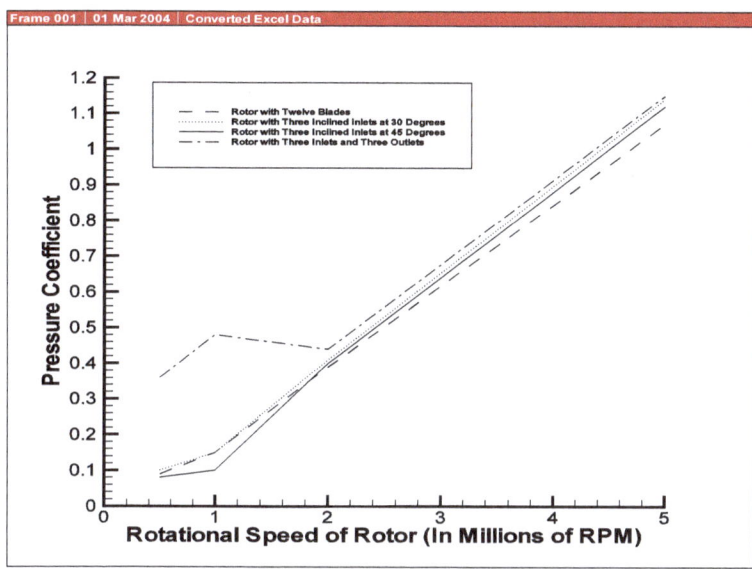

Fig. 2.7 Variation of pressure coefficient with rotating speed of the rotor for different geometries of the rotor such as rotor with twelve blades, rotor with three inlets inclined at an angle of 30°, rotor with three inlets inclined at an angle of 45°, and rotor with three inlets and three outlets. Reproduced with permission. Copyright retained by Inderscience Publishers

Figure 2.7 shows the variation of pressure coefficient values for rotor with twelve blades, rotor with inlets, inclined at an angle of 30°, rotor with inlets, inclined at an angle of 45° and rotor with three inlets and three outlets for different rotating speeds of the rotor. From Fig. 2.7 it could be said that as the rotating speed of the rotor increases, pressure coefficient increases and the pressure coefficient values are almost the same for all geometries with the increase in rotational speed similar to the previous case.

From Table 2.2 It can be concluded that for both half-million rpm and one million rpm rotating speed of rotor, rotor with three inlets inclined at an angle of 45° is the optimum design compared to the other geometries of the rotor considered after modifying the rotor geometry and for two million and five million rpm rotational speed of the rotor, the rotor with twelve blades is the optimum design of rotor for high-speed spindles compared to the other geometries of the rotor. Rotor geometries considered for numerical simulations of high-speed spindles were the two-stage rotor and its modified geometry. The values of pressure coefficients obtained for different rotational speeds of these rotors are given in Table 2.3. The number of control volumes associated with the numerical simulations of two-stage rotor and its modified geometry were 522,046 and 521,614, respectively. The maximum number of iterations considered for convergence control was 200 and the residual target for convergence criteria was specified as 1e-6 for the geometries. From the table, it could be observed that the pressure coefficient values obtained for the numerical simulations of two-stage rotor and its modified geometry are

Table 2.3 Pressure coefficient values for rotor geometries two-stage rotor and its modified geometry

Geometry	Half-million	One Million	Two Million	Five Million
Two-stage rotor	1.17	1.19	1.20	1.46
Modified geometry of two-stage rotor	1.21	1.26	1.35	1.65

Reproduced with permission. Copyright retained by Inderscience Publishers

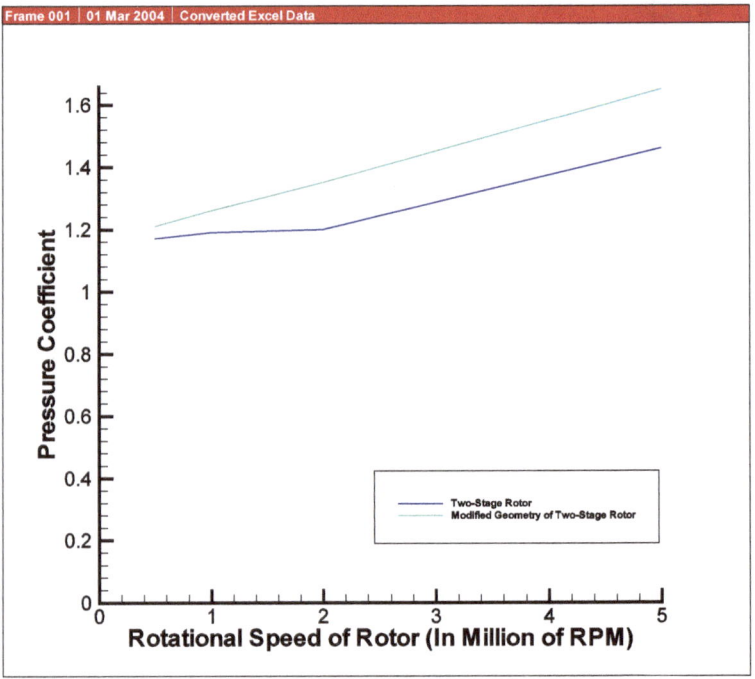

Fig. 2.8 Variation of pressure coefficient with rotational speed for two-stage rotor and its modified geometry. Reproduced with permission. Copyright retained by Inderscience Publishers

very high compared to other geometries of rotor for half-million, one million and two million rpm rotational speeds. The pressure coefficient was found to be almost the same with other geometries of the rotor for five million rpm. Figure 2.8 shows the variation of pressure coefficient with rotational speed of rotor for two-stage rotor and its modified geometry.

From all the geometries discussed above, it can be concluded that for half-million rpm rotational speed of the rotor, rotor with 90° blade angle is the optimum geometry for rotor and for one million rpm rotating speed of the rotor, rotor with three inlets inclined at an angle of 45° is the optimum design compared to other geometries of the rotor. For two million and five million rpm rotational speed of the rotor, rotor with twelve blades is the optimum design of rotor for high-speed spindle compared to the other geometries of the rotor.

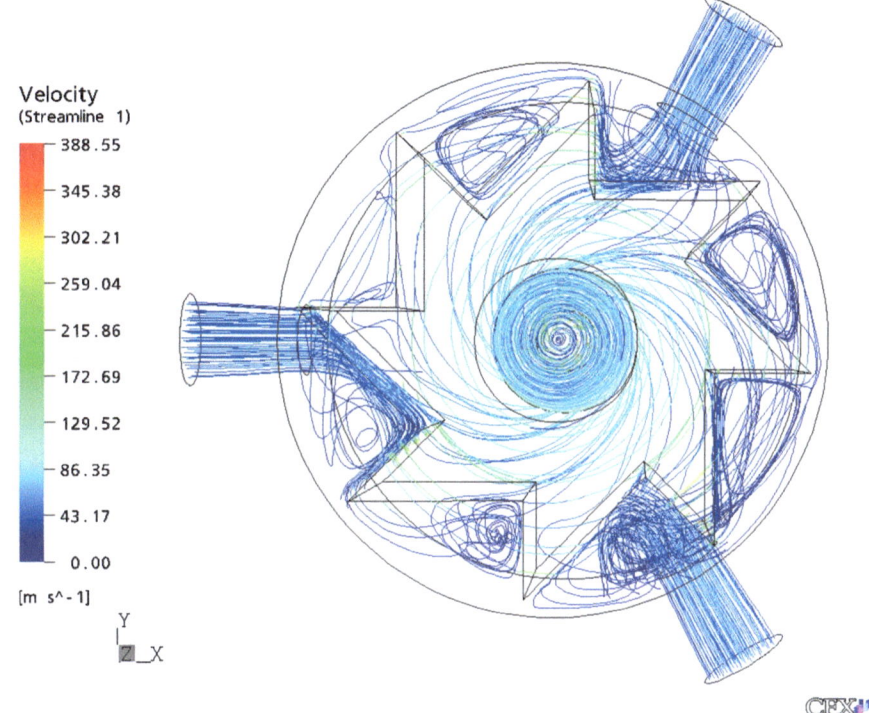

Fig. 2.9 Streamlines describing the flow pattern for rotor with 90° blade angle at one million rpm rotational speed of rotor. Reproduced with permission. Copyright retained by Inderscience Publishers

Flow Topology and Pressure Variations

Flow topology and pressure variation of rotor geometries such as rotor with 90° blade angle, rotor three inlets inclined at an angle of 45°, and two-stage rotor are described.

Rotor with 90° Blade Angle

Flow topology and pressure variation of this geometry will be explained as the pressure coefficient of this geometry was found to be lower than for the other geometries of rotors. The rotational speed of the rotor was specified as one million (1,000,000) RPM.

Flow Topology: Fig. 2.9 shows the flow pattern in rotor with 90° blade angle. Air enters into the housing from three inlets with a pressure of 60 psi and exists from outlet at a pressure of zero. The three inlets are normal to the surface of the housing and the outlet is at the centre of the housing. Air entering from the inlets impinges on the rotor surfaces and flows inside of housing. At the hollow part of the rotor, the fluid is swirled and at the outlet the velocity increases due to the sudden contraction of the surface area at the outlet.

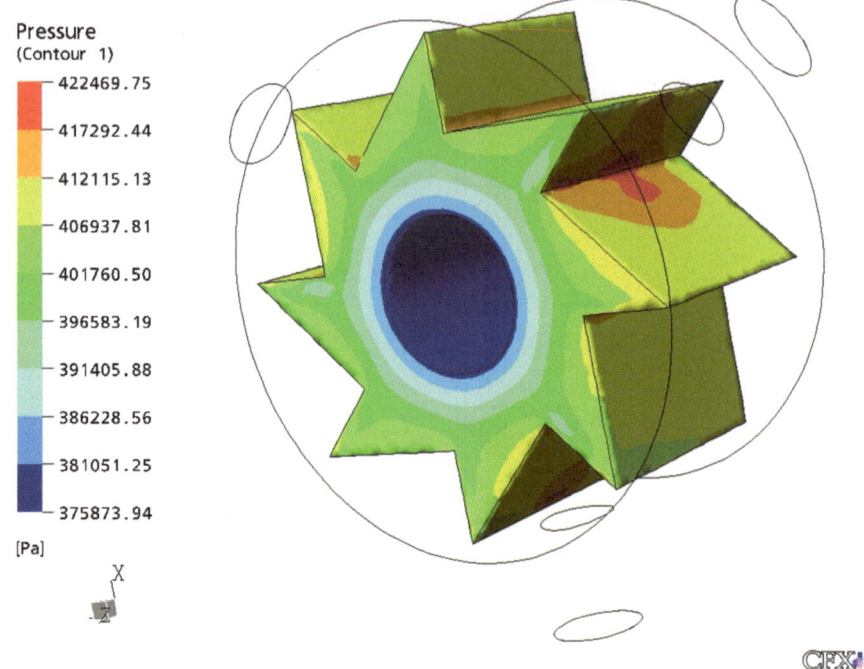

Fig. 2.10 Static pressure distribution on the rotor with 90° blade angle at one million rpm rotational speed of rotor. Reproduced with permission. Copyright retained by Inderscience Publishers

Pressure Variation: Fig. 2.10 describes the pressure variation on the surfaces of the rotor. From the figure it can be observed that the regions of the rotor, where the fluid impinges directly are having maximum pressure due to the stagnation of the fluid compared to the other regions of the rotor. At the stagnation point the total kinetic energy of the fluid is converted into pressure energy, so the maximum pressure occurs at the stagnation point of the fluid. The pressure coefficient on the rotor can be determined from the Eq. 2.9 by substituting values of maximum pressure and minimum pressure on the rotor as shown in Fig. 2.10 and inlet pressure (60 Psi).

Rotor with Three Inlets Inclined at 45°

Flow topology and pressure variation on the rotor with three inlets, inclined at an angle of 45° with z-axis is described. The rotational speed of the rotor was specified as half-million (500,000) rpm.

Flow Pattern: Flow pattern in the geometry can be seen in the Fig. 2.11. Air is entering into the housing from three inlets with a pressure of 60 psi and exists from outlet at a pressure of zero. Fluid entering from three inlets impinges on the rotor blades directly, unlike the previous case (Fig. 2.9), where flow from the inlets is not impinging directly on the rotor blades, but impinging in between the rotor blades. Figure 2.11 clearly shows the impingement of the fluid on the rotor blades. At the hollow part of the rotor, the fluid is swirled and at the outlet the velocity

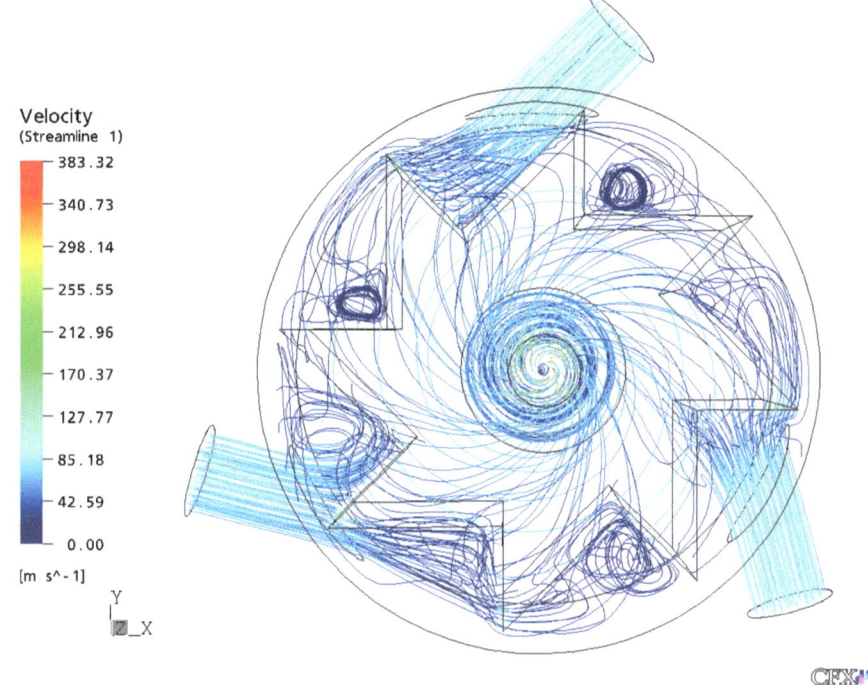

Fig. 2.11 Stream lines describing the flow pattern for rotor with three inclined inlets at an angle of 45° for half-million rpm rotational speed. Reproduced with permission. Copyright retained by Inderscience Publishers

increases due to the sudden contraction of the surface area at the outlet. This forms a vortex flow at the outlet.

Pressure Variation: Fig. 2.12 shows the pressure variation of rotor, along with housing, outlet and three inlets on a plane parallel to X–Y plane, at a distance of 0.08675 inches from the centre, in the z-direction. From the figure it could be seen that the regions of the rotor, where the fluid flow directly impinges are having maximum pressure due to stagnation of fluid when compared to other regions. At the outlet, which is at the centre of the housing pressure is found to be minimum. Figure 2.13 shows the three-dimensional view of the pressure variation on the rotor surfaces. The legend shows the pressure values at different surfaces of the rotor.

Two-Stage Rotor

Flow topology and pressure variation of two-stage rotor is described.

Flow Topology: The inlet static pressure was considered to be 60 psi and outlet static pressure as zero. Rotor's rotational speed was specified as two million rpm (2,000,000). Flow topology is illustrated in Fig. 2.14. Air entering from the upper part of the housing (through inlets 1 and 2) is diverted by rotor blades. One part slides through the upper part of the housing and exits to the atmosphere through outlets 3 and 4, and the other flows over the rotor blade into the housing and exits

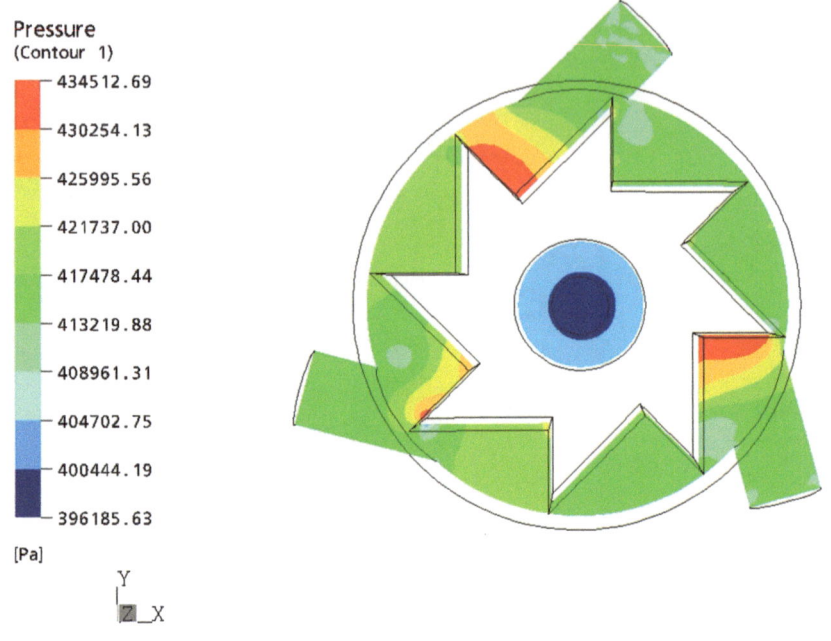

Fig. 2.12 Static pressure variation on the spindle parallel to the X–Y plane for half-million rpm rotational speed. Reproduced with permission. Copyright retained by Inderscience Publishers

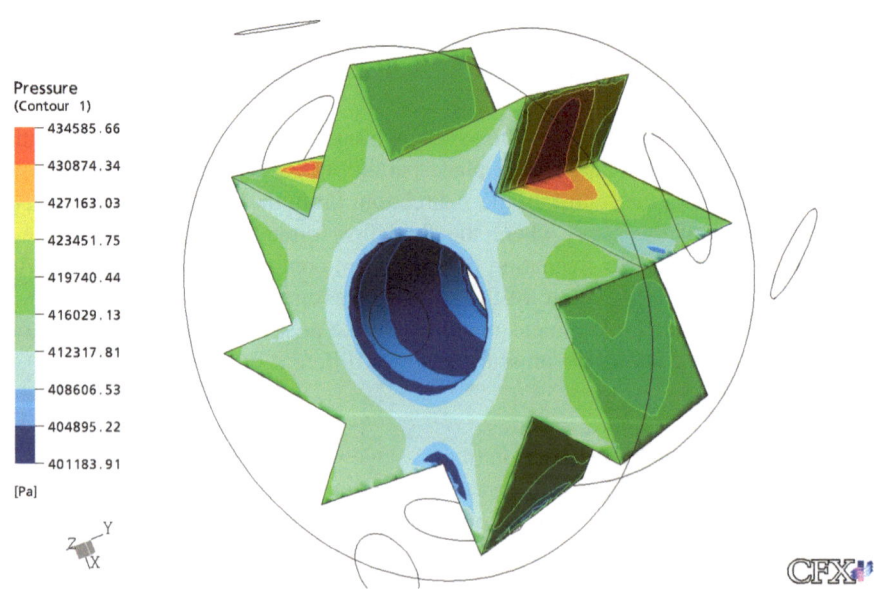

Fig. 2.13 Static pressure distribution on the rotor with three inclined inlets at an angle of 45° for half-million rpm rotational speed of rotor. Reproduced with permission. Copyright retained by Inderscience Publishers

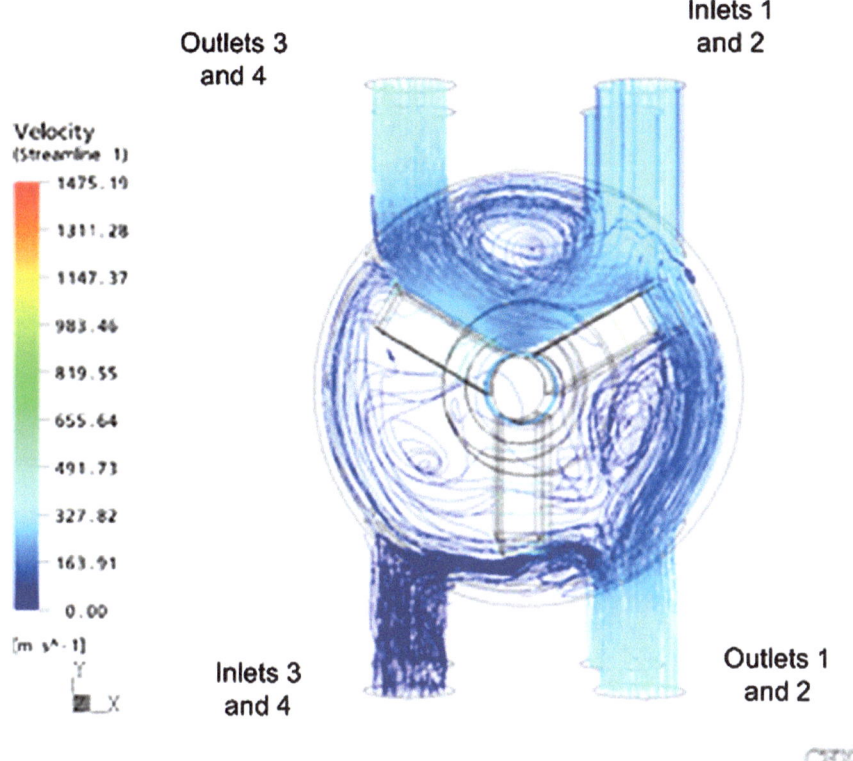

Fig. 2.14 Stream lines describing the flow pattern of two-stage rotor for two million rpm rotational speed of rotor. Reproduced with permission. Copyright retained by Inderscience Publishers

to the atmosphere through the outlets 1 and 2. Fluid entering from the lower part of the housing (through inlets 3 and 4) is bifurcated and most of the fluid builds a re-circulation zone in the lower part of the housing and exits to the atmosphere through outlets 1 and 2 and the other part flows into the housing, and exits to the atmosphere through outlets 3 and 4. From Figs. 2.9, 2.11 and 2.14, it could be observed that the maximum values of velocity of fluid are different, as the rotational speeds of the rotor are different. For rotor with high rotational speed, the maximum velocity is high compared to other rotors.

Pressure Variation: In Fig. 2.15, static pressure distribution of spindle geometry obtained from the numerical simulation of rotor is shown parallel to the X–Y plane at a distance of 3 mm from one side of the rotor (and at a distance of 11 mm from the other side of the rotor) in the z-direction. Air entering from the lower part of the housing impinges directly on the rotor blade surface and the total kinetic energy of the fluid gets converted into the pressure energy due to the stagnation of the fluid. Therefore, the maximum pressure is observed in the vicinity of the leading edge of the rotor blades, which bifurcate the inlet stream entering from the lower part of the housing. Figure 2.16 shows the static pressure distribution on the rotor. The pressure distribution was found to be symmetric on the blades of the rotor.

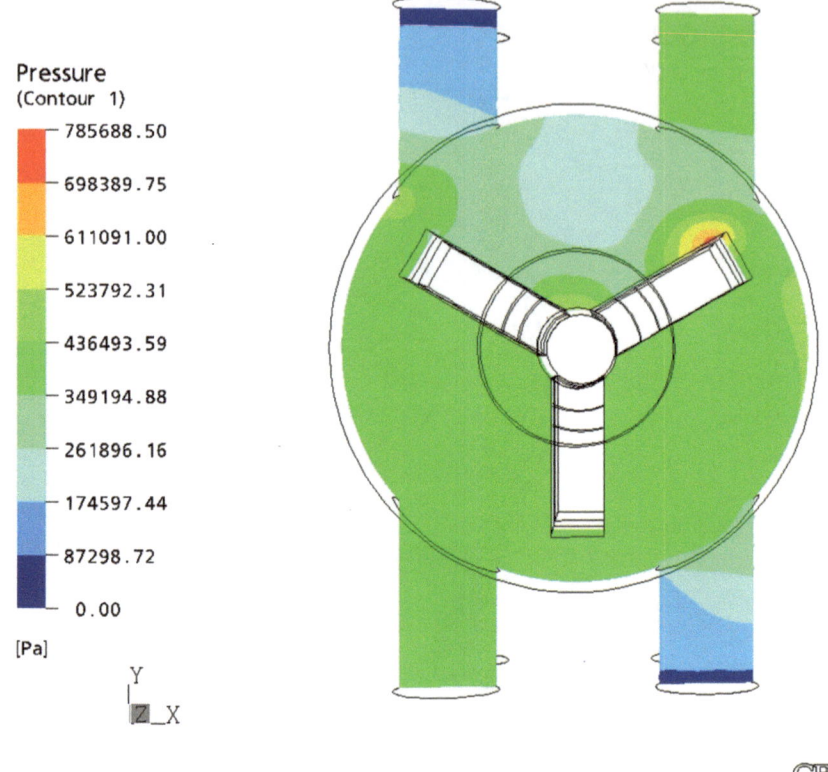

Fig. 2.15 Static pressure contours parallel to X–Y plane for two million rpm rotational speed of rotor. Reproduced with permission. Copyright retained by Inderscience Publishers

Mach Number

Mach number plays an important role for high-speed flows. Mach number on the surfaces of the rotor was found using FLUENT^{TM}. For all the geometries of the rotor at the rotational speeds of half-million rpm, one million and two million rpm, the Mach number is within the subsonic regime, with only a few spots exceeding a Mach number of 1. For rotating speed below two million, the subsonic assumption is well justified. However, for higher rpm, the transonic and supersonic flow regimes need to be taken into consideration.

Using Fluent: Mach number on the surfaces of the rotor was determined using CFD software. FLUENT^{TM} needs the user to specify the flow as subsonic, or supersonic flow as input. Fluent does not require that specification before the calculation. The regime of the flow (subsonic, transonic, or supersonic) comes as part of the solution. The geometry of the rotor with three inlets, inclined at 45° with three inlets and a housing was created and the mesh was created. Volume elements considered for the mesh were approximately 600,000. The ideal gas, air, was considered as the fluid and the k-epsilon

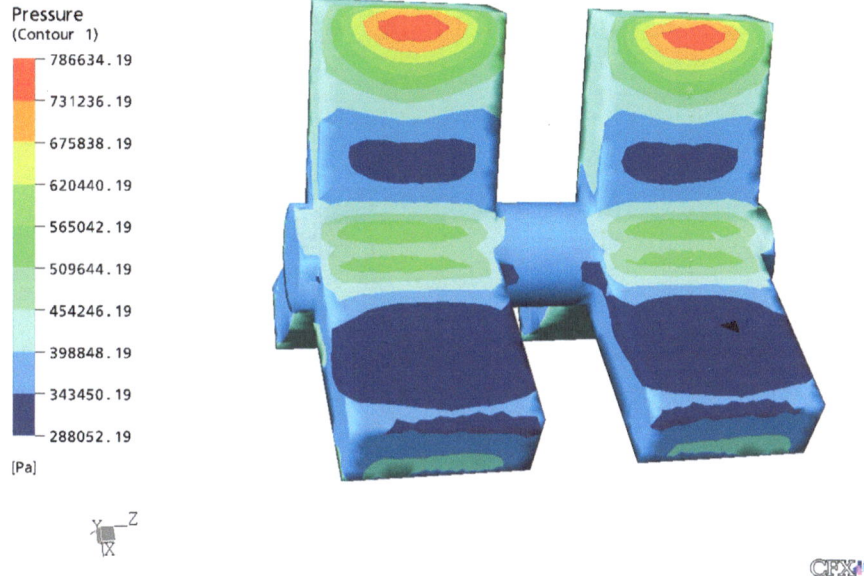

Fig. 2.16 Static pressure distribution on the two-stage rotor for two million rpm rotational speed of rotor. Reproduced with permission. Copyright retained by Inderscience Publishers

turbulence method was used for modelling turbulent flow. An inlet boundary condition, with a pressure of 60 psi was considered at three inlets. At the outlet, outlet boundary condition with a pressure of zero was considered. The rotor was defined as a moving wall with no-slip boundary condition, with an angular speed of two million rpm along its axis. For the remaining surfaces of the geometry, a default no-slip stationary wall boundary condition was applied.

For this compressible flow, a coupled implicit scheme was considered for the solution. The maximum number of iterations for the convergence control for the solver was specified as 4,500 and the target residual for the convergence criterion was specified as 1e-6. The governing equations such as continuity, momentum, energy equations, and equation of state were solved by FLUENT^{TM} and the Mach number at various zones was determined. Figure 2.17 shows the variation of Mach number at three inlets, outlet, and on the rotor surface for the rotational speed of the rotor of two million rpm. From the figure it can be observed that the maximum value of Mach number is 0.96 and the minimum value is 0.01. At the inlets and on the surfaces of the rotor, the Mach number is significantly less than 1 and at the outlet, the value is 0.96. From the results it can be said that only at the outlet, the flow is transonic, as the velocity of the flow is maximum at the outlet because of zero pressure condition. At the inlets and on the rotor surfaces, the flow is found to be subsonic.

An experimental high-speed spindle has been manufactured and is shown in Fig. 2.18. Further experiments are planned that will measure the velocity of the rotor shown in Fig. 2.18 at various inlet air pressures.

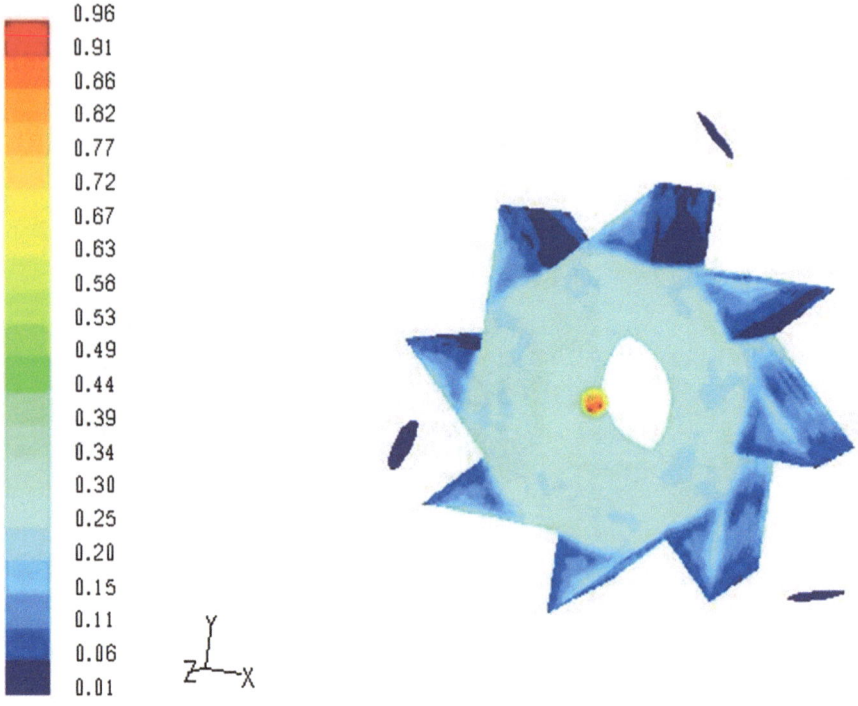

Fig. 2.17 Mach number contours for rotor with three inlets, inclined at an angle of 45° for two million rpm rotational speed of rotor. Reproduced with permission. Copyright retained by Inderscience Publishers

Fig. 2.18 Construction of the prototype three-blade rotor. Reproduced with permission. Copyright retained by Inderscience Publishers

Fig. 2.19 Schematic of
the rotor. Reproduced with
permission. Copyright
retained by Inderscience
Publishers

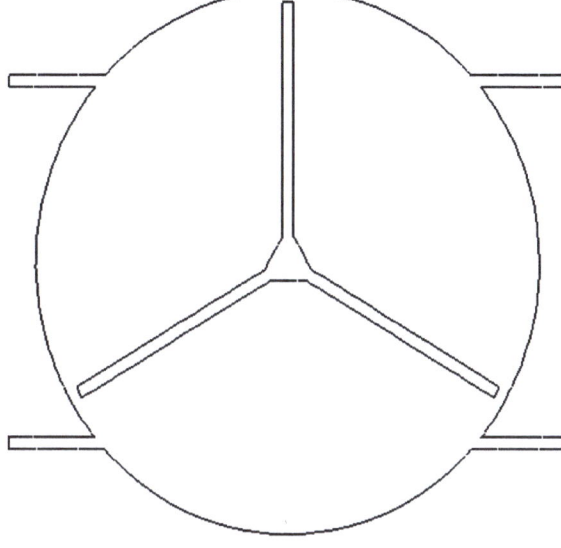

Fig. 2.20 Mesh of
the experimental rotor.
Reproduced with permission.
Copyright retained by
Inderscience Publishers

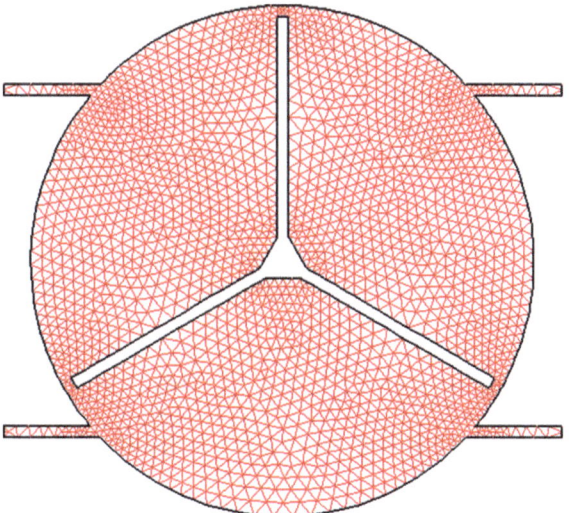

The experimental spindle has been modelled using FLUENT^{TM} to determine the pressures and velocities. The schematic of the spindle is shown in Fig. 2.19.

The grid for this geometry is shown in Fig. 2.20. The grid has been generated in GAMBIT^{TM}. The grid is made of total 2,200 points and triangular mesh elements. The grid is clustered near the tip clearance and near the hub. The inlet to the domain is based on the stagnation pressure and temperature inlet BC while at the outlet the back-pressure of suction is applied. The fluid domain is divided into 2 parts: static fluid and rotating fluid. The static fluid is the fluid portion in the straight ducts of the inlet and outlet where the fluid is not rotating. The rotating

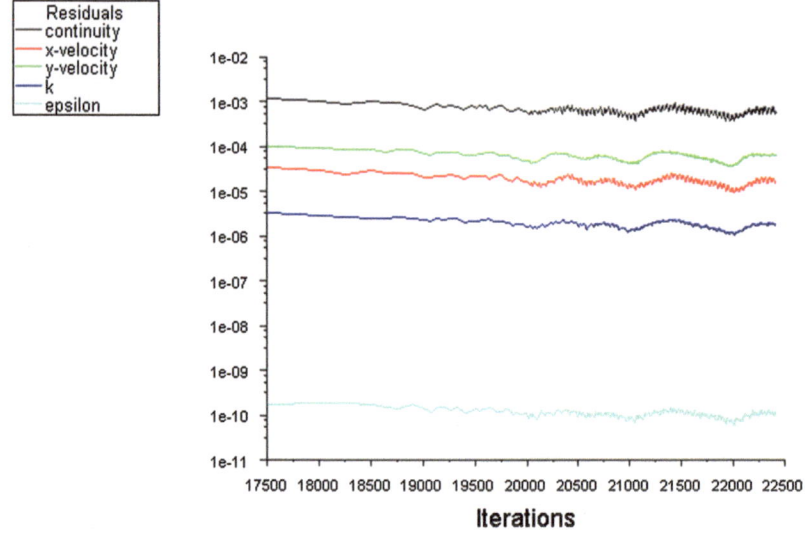

Fig. 2.21 The residual plot of the CFD analysis of the experimental rotor. Reproduced with permission. Copyright retained by Inderscience Publishers

fluid is the fluid surrounding the blades and enclosed in the casing. The angular rotation of half a million rpm is to be applied to the rotating fluid which is explained later.

The residual plot of the rotor is seen in Fig. 2.21, which shows that the solution has not converged. Further analysis is required to produce fully converged results.

The static pressure contours are shown in Fig. 2.22 for the experimental rotor.

The x- and y-velocity contours are shown in Fig. 2.23, whilst Fig. 2.24 shows the total pressure exerted with accompanying flow lines.

The magnitude of velocities and pressure are too high and thus indicates us to re-evaluate our boundary conditions used in the problem. Additional computations can be done with a much denser grid to capture the unsteady dynamics of the problem. Flow topology and pressure variation of different geometries has been described lucidly. Pressure coefficient values of all geometries were given and the variation of pressure coefficient value with the rotating speed of rotor was presented. It can be observed that changes in inlet, outlet, and rotor geometries affect the pressure coefficient significantly in air turbine powered dental hand pieces.

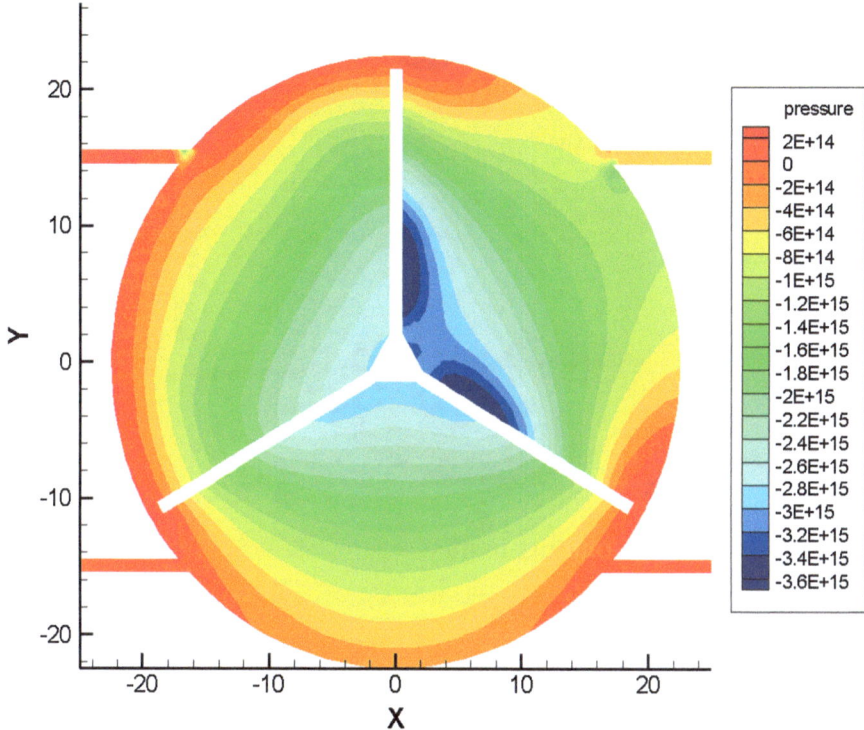

Fig.. 2.22 Static pressure contours of the experimental rotors. Reproduced with permission. Copyright retained by Inderscience Publishers

2.5 Clinical Environment

Conditions in the surgery are different from the laboratory (Silvaggio and Hicks 1997; Borges et al. 1999). The purpose of the coating process is to lengthen the working life of a bur. They will then be available for use on multiple patients, and therefore undergo sterilisation processes. Sterilisation procedures may increase the risk of instrument fracture. It also has the potential to be corrosive and may affect the cutting surface or coating of the bur.

The main sterilisation methods are autoclaves and hot-air sterilisation. The autoclave is the best method of sterilisation for dental instrumentation. The instruments are heated in steam under pressure until conditions are reached which are beyond the survival of any microorganism. Modern autoclaves use electronic programming and monitoring of the condition in the steam chamber in order to ensure that sterilisation is achieved. Hot-air sterilisations are basically simple electric ovens. Hot dry air sterilises effectively but takes much longer than wet heat.

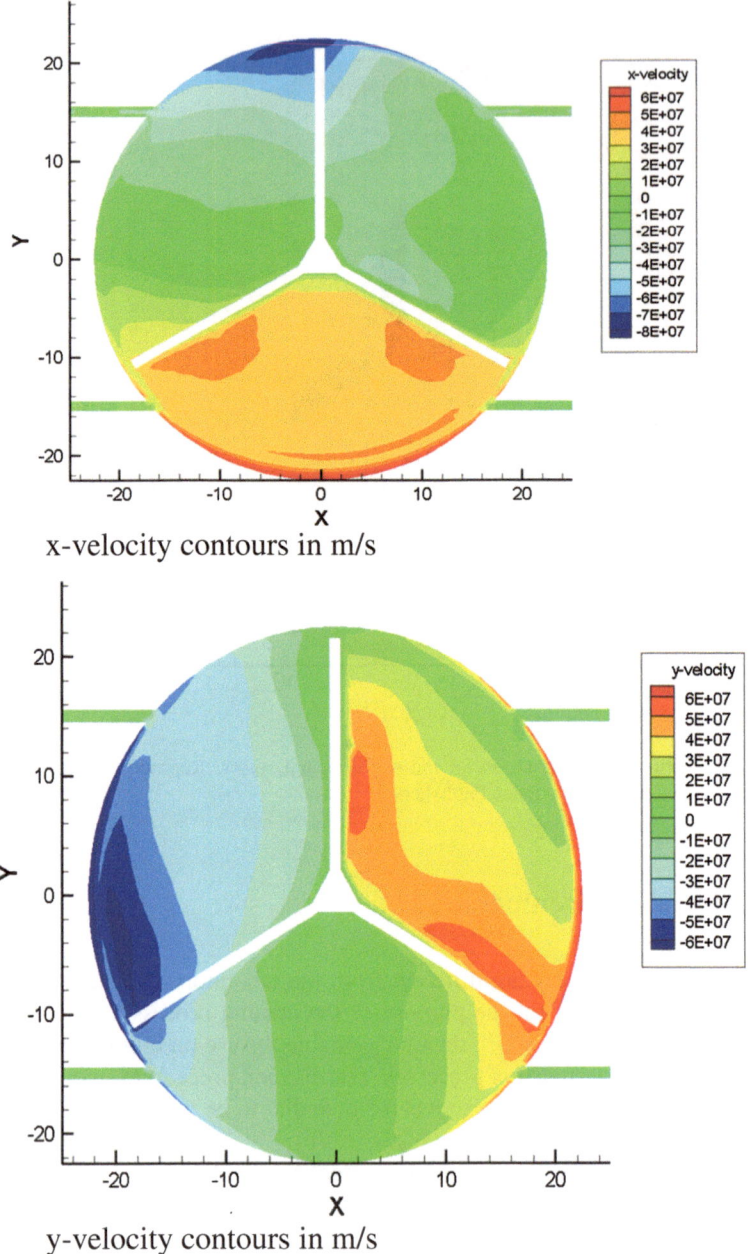

x-velocity contours in m/s

y-velocity contours in m/s

Fig. 2.23 Velocity contours for the experimental rotor. Reproduced with permission. Copyright retained by Inderscience Publishers

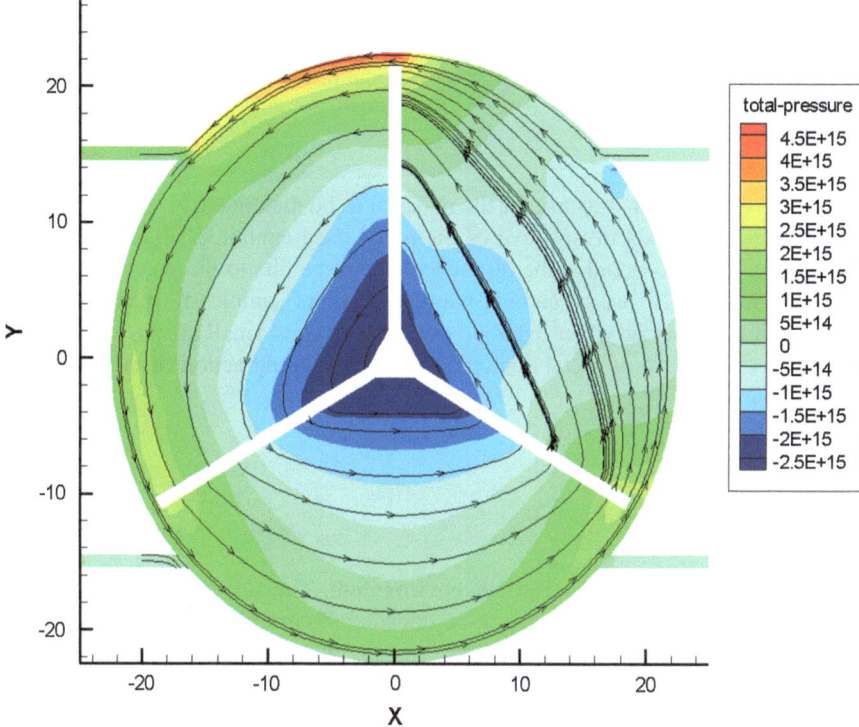

Fig. 2.24   Total pressure in Pa with accompanying flow streamlines for the experimental rotor. Reproduced with permission. Copyright retained by Inderscience Publishers

2.6 Machining Materials with Dental Burs

Wide ranges of tests have been carried out on dental burs. Surprisingly few of these consider the cutting surface and performance of the bur. Many use an approach addressing the physical properties of instrumentation under conditions that are not similar to clinical operating conditions. Those that do simulate clinical conditions tend to be specific with respect to the area of interest and thus the results have very limited relevance outside of the direct application. Also there is no standard covering the testing of burs regarding their performance. The vast majority of the research carried out has used different machine settings, materials methodology and result measurement. This makes it hard to assess the significance of the individual findings and nearly impossible to obtain quantifiable comparison between reports. This research therefore has no fixed guidelines to follow and is entering a field in which there is little background knowledge available.

The most applicable study for this research work is an investigation carried out by Watanabe et al. (2000). They studied cutting performances of air turbine burs on a selection of materials including a range of titanium alloys. They evaluated the machinability after constant cutting for 5 s. The data was adjusted to account for the density of each metal used. These researchers found for the carbide fissure bur that machinability (**M**) was inversely proportional to the metal hardness (H).

$$M = \frac{k}{H} \tag{2.10}$$

where, k is a empirical constant for the material to be machined.

The diamond points irrespective of the metal used exhibited similar machining efficiencies. It was concluded the differences might be due to the way in which the burs etched. With the use of carbide burs the titanium samples formed accordion-shaped chips. However, when diamond points were used small shapes was formed depending on the size of metal demonstrating the grinding action of the embedded diamond particles.

2.7 Tooth Materials

The mouth is a potentially hazardous environment. The selection of materials for any given application must therefore be considered carefully (McCabe and Walls 2009). Within the oral cavity conditions vary in an irregular manor. For example the variation of temperature in a day may be between 32 and 37 °C depending on whether the mouth is open or closed. The ingestion of hot or cold food or drink however, can extend this temperature range to between 0 and 70 °C. Other important variables include acidity, humidity and tooth impact loading.

2.7.1 Human Tooth

Teeth can be considered as hard, calcified structures. They are attached to the upper and lower jaw, used primarily for chewing food. Each tooth consists of a crown, protruding into the mouth and a root, extending down into the socket as shown in Fig. 2.25, where the junction of the crown and root is the neck of the tooth.

Enamel is the white cap on the crown of the tooth. It is the hardest substance in the body: a calcified material with a very small organic content (less than 4 %). Dentine lines the deep surface of the enamel in the crown, and continues down into the root, surrounding a central pulp cavity. Dentine is like enamel: a calcified matrix, but softer, with some 27 % organic matters, which is mainly collagen. It is believed to house very fine sensory nerve endings.

The dentine is covered on its superficial surface by a layer of cement, yet another calcified tissue going from the neck to the root of the tooth. It has a higher organic content, 50 % by weight. Resembling bone in composition, it is the tissue that receives the collagen fibres of the periodontal membrane, which anchor the tooth to the walls of the socket.

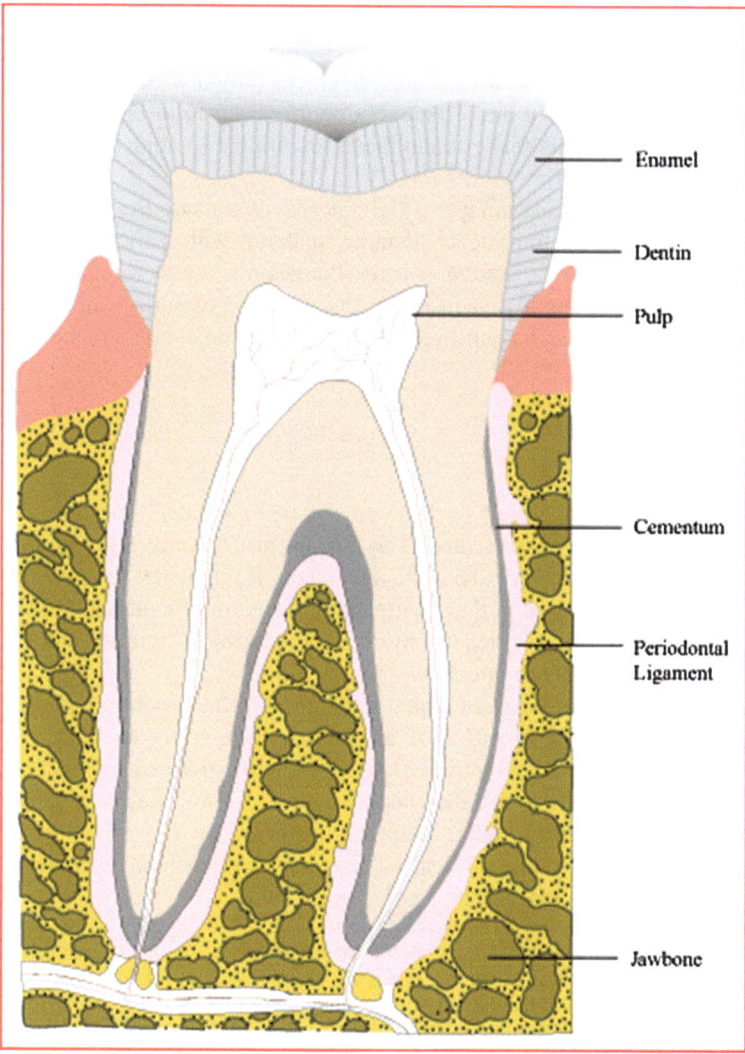

Fig. 2.25 Structure of a human tooth. http://www.vaughns-1-pages.com/local/sunnyvale

2.7.2 Artificial Tooth Materials

Artificial teeth are required to withstand the conditions and loads of normal teeth. The artificial teeth should be strong and tough in order to resist fracture. They should also be hard enough to resist abrasive forces in the mouth and during cleaning, but should allow grinding with a dental bur so adjustments to the occlusion can be made by the dentist at the chair side. The materials most widely used for the manufacture of artificial teeth are acrylic resin and porcelain (McCabe and Walls 2009).

Acrylic resin artificial teeth are produced in reusable moulds by dough or injection moulding. The resins used are highly cross-linked to produce teeth, which are resistant to crazing (when a series of surface cracks have a weakening effect on the whole piece). Acrylic resin has a relatively poor resistance to fatigue fracture and poor impact strength; this makes them more susceptible to abrasion. Acrylic resin is a good thermal insulator.

Porcelain teeth are made in layers. The inner layer is made from a fairly opaque 'core' material. A more translucent 'dentine' material with a translucent 'enamel' porcelain coating forms the outermost layer. Porcelain is a very rigid, hard and brittle material whose strength is reduced by the presence of surface irregularities or internal voids and this can be improved by adding powered alumina to the porcelain.

2.8 Conclusions

In this chapter a brief overview of the dental tools, materials, conditions and operating conditions has been presented. The chapter also describes the oral environment where the dentists work. The size and shape of the dental tool is dictated by where the tool is to be used in the mouth and the amount of material removed. The choice of material is highly dependent on the operating conditions and the environment in which dental bur operates.

When considering high speed dental handpieces, flow topologies and pressure variations of different geometries of air turbine have been described. Pressure coefficient values of all geometries of turbines were given and the variation of pressure coefficient value with the rotating speed of rotor was presented. It can be observed that changes in inlet, outlet, and rotor geometries affect the pressure coefficient significantly. A great deal of research and development is required in the area of high speed dental handpieces.

Acknowledgments The authors are grateful to Inderscience for allowing the authors to reproduce material published in the International Journal of Nano and Biomaterials, 2009, Volume 2, Number 6, p. 505. Inderscience retains copyright of the material used in this chapter. The authors thank their graduate students for contributing to this chapter in helping to formulate the numerical models.

References

Ahmed W et al (2004) Chemical vapour deposition of diamond films onto tungsten carbide dental burs. Tribol Int 37(11–12):957–964

Borges CFM et al (1999) Dental diamond burs made with a new technology. J Prosthet Dent 82(1):73–79

Carvalho CAR et al (2007) The use of CVD diamond burs for ultraconservative cavity preparations: a report of two cases. J Esthetic Restorative Dent 19(1):19–29

Dietz DB et al (2000) Effect of rotational speed on the breakage of nickel-titanium rotary files. J Endod 26(2):68–71

Jian XG et al (2004) Study on the effects of substrate grain size on diamond thin films deposited on tungsten carbide substrates. Key Eng Mater 274:1137–1142

McCabe JF, Walls A (2009) Applied dental materials. Wiley, London

O'Brien WJ (1997) Dental materials and their selection. Quintessence Publication, Chicago

Polini R et al (2004) Cutting force and wear evaluation in peripheral milling by CVD diamond dental tools. Thin Solid Films 469:161–166

Seely RR et al (2007) Essentials of anatomy and physiology. McGraw-Hill, New York

Sein H et al (2003) Stress distribution in diamond films grown on cemented WC–Co dental burs using modified hot-filament CVD. Surf Coat Technol 163:196–202

Sein H et al (2004a) Enhancing nucleation density and adhesion of polycrystalline diamond films deposited by HFCVD using surface treatments on Co cemented tungsten carbide. Diam Relat Mater 13(4–8):610–615

Sein H et al (2004b) Performance and characterisation of CVD diamond coated, sintered diamond and WC–Co cutting tools for dental and micromachining applications. Thin Solid Films 447:455–461

Sein H et al (2002) Application of diamond coatings onto small dental tools. Diam Relat Mater 11(3–6):731–735

Sein H et al (2006) Comparative investigation of smooth polycrystalline diamond films on dental burs by chemical vapor deposition. J Mater Eng Perform 15(2):195–200

Siegel SC, von Fraunhofer JA (1999) Irrigating solution and pressure effects on tooth sectioning with surgical burs. Oral Surg Oral Med Oral Pathol Oral Radiol Endod 87(5):552–556

Silvaggio J, Hicks ML (1997) Effect of heat sterilization on the torsional properties of rotary nickel-titanium endodontic files. J Endod 23(12):731–734

Trava-Airoldi VJ et al (1996a) CVD diamond burs—development and applications. Diam Relat Mater 5(6):857–860

Trava-Airoldi VJ et al (1996b) Development of chemical vapor deposition diamond burs using hot filament. Rev Sci Instrum 67(5):1993–1995

Watanabe I et al (2000) Cutting efficiency of air-turbine burs on cast titanium and dental casting alloys. Dent Mater 16(6):420–425

Yared GM et al (2001) Influence of rotational speed, torque and operator's proficiency on ProFile failures. Int Endod J 34(1):47–53

Chapter 3
Growth and Application of Diamond Thin Films

Abstract The chapter describes the growth and application of thin films. The development of the hot filament CVD process is described in terms of surface pre-treatments and CVD equipment preparation. The chapter then focuses on how metal cutting theory can be applied to the machining of bone which is an essential procedure in dental practice.

Keyword Diamond · Chemical vapour deposition · Bone · Machining · Dental practice

3.1 Introduction

The modified hot filament chemical vapour deposition (HFCVD) system has been described. The initial testing of the system in terms of filament characteristics, power considerations and ultimately the deposition experiments were carried out in order to optimise the substrate pre-treatments and the experimental CVD conditions for the diamond growth.

Specific experimental conditions and processes are outlined within the relevant chapters.

3.2 CVD System

A HFCVD system was modified to enable deposition onto complex and flat shaped different substrates and for substrate biasing experiments carried out during the deposition of hard carbon coatings. Figure 3.1 shows the HFCVD and modified vertical filament (VF) CVD system arrangement (Sein et al. 2002a, b).

W. Ahmed et al., *Chemical Vapour Deposition of Diamond for Dental Tools and Burs*, SpringerBriefs in Materials, DOI: 10.1007/978-3-319-00648-2_3, © The Author(s) 2014

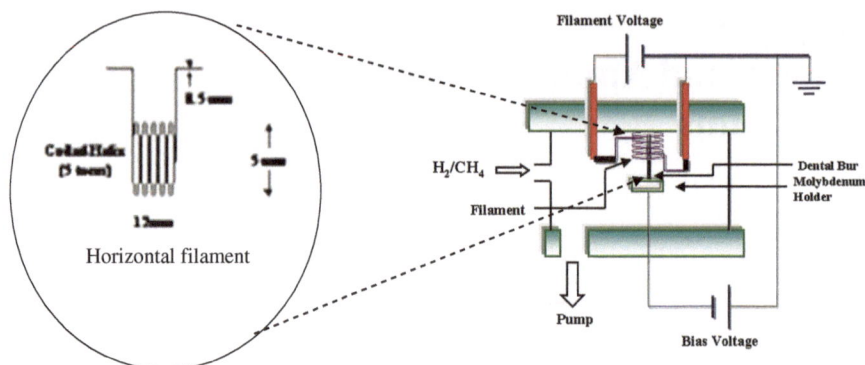

Fig. 3.1   Deposition chamber employing the modified VFCVD system (Sein et al. 2002a, b)

The system consists of the following components:

- DC power supply (HP, 0–40 V and 0–30 A)
- Cylindrical stainless steel reaction chamber
- Mechanical vacuum pump
- Mass flow controllers (MFC)
- Mass flow control read out unit (unit instrument: URS.100.5)
- Water-cooling system (built-in within the chamber)
- Bourdon pressure measurement gauge
- K-type thermocouple
- Two colour optical pyrometer
- Molybdenum substrate holder
- Filament arrangement for different substrates (99.9 % pure Tantalum wire of 0.5 mm diameter)

The modified hot filament chemical vapour deposition (VFCVD) system enabled the substrates to be negatively biased with respect to the filament. In the cases where the substrates were negatively biased, prior to substrate biasing the filament was pre-carburised for 15 min under standard CVD conditions using 3 % methane concentration to avoid tantalum contamination.

The system can be divided into three main components: the gas supply system; CVD process chamber and pumping system. Firstly, the reactant gases, typically an excess of H_2, CH_4 and, if required, Ar are delivered to a pre-mixing manifold via separate mass flow controllers (MFC Unit Instruments Ltd) before being admitted to the process chamber. This arrangement, allows the various hydrogen, methane and argon ratios to be set independently up to a total flow-rate of 200 sccm.

The water-cooled stainless steel reaction chamber houses a removable substrate-filament assembly located in the plane of a viewing glass window. Gas activation is achieved by a single 0.5 mm diameter tantalum filament of 100 mm total length, which was coiled and supported by stainless steel rods leading to an isolated power feed made of copper rod. This ensured that maximum power dissipation was restricted only to the filament section of the complete heating circuit. The filament in turn, was powered by a 2.5 kW DC power supply. About 5 mm below the filament, flat silicon {100} substrates (5 mm × 5 mm) were mounted on a 20 mm × 20 mm molybdenum holder. Figure 3.1 shows that the horizontal and vertical filament arrangements are interchangeable depending on the application for which it is used.

The micro drill and dental bur substrates were mounted vertically on the molybdenum substrate holder. The filament was also mounted vertically with the dental bur or microdrill held within the filament coils, as opposed to the horizontal position used in the conventional HFCVD system. To ensure uniform coating the substrate was positioned centrally and coaxially within the coils of the filament.

The substrate temperature was measured with an insulated type K thermocouple embedded in the substrate holder and lying flush with the substrate. The whole of the substrate mounting and heating assembly could be translated with respect to the filament within 0.1 mm, allowing both accurate and reproducible sample positioning. Pumping was achieved via a single-stage rotary pump, which could achieve a base pressure of 1×10^{-3} Torr determined by a Pirani gauge located at the back of the chamber. The total process pressure, typically of the order of 20 Torr (2.66 kPa) was measured using a Bourdon pressure gauge.

To optimise the VFCVD process to grow high quality diamond films, the deposition conditions were carefully controlled. The parameters influencing gas activation are the type of filament material (before and after carburisation), filament temperature and filament geometry, but also the filament-substrate distance. Since all these factors influence the concentration of activated species reaching the substrate surface, they must be optimised. The initial deposition experiments were carried out to characterise the VFCVD system and hence determine the optimum conditions for diamond synthesis had been established. Deposition conditions were modified along with the design and process parameters to obtain polycrystalline diamond films similar to those obtained previously (Wang et al. 2000).

In diamond HFCVD system, the mechanism of heat transfer is prominent and is also driving force in chamber. In addition to radiation, convection and conduction heat transfer is also achieved via atomic hydrogen since the formation of atomic hydrogen at or near the filament surface is highly endothermic. Atomic hydrogen readily recombines on solid surfaces to form molecular hydrogen with the recombination reaction being highly exothermic. Thus, atomic hydrogen acts as a carrier of heat from the filament to the growth surface. Since the quality, morphology and defect density of diamond films are sensitive to temperature, a uniform substrate temperature is crucial for the deposition of diamond films with uniform properties. Therefore, the heat transfer to the substrate and the resulting substrate temperature distribution are important considerations in the design of reactors for coating large areas and deposition of complex shaped substrates.

Table 3.1 Trace elements in
the tantalum wire (Datasheet:
Sigma Aldrich Catalogue
(2001)

Element	Concentration (ppm)
Al	120
Cu	85
Sn	80
Ni	35
Cr	20
V	15
Mg	1

3.2.1 Filament Set-Up for Flat substrate (Conventional HFCVD System)

The characteristics of the filament required for CVD of diamond films were
described in Chap. 1. Filaments used in this work were made from 0.5 mm diameter tantalum wire by wrapping it around a 4.5 mm diameter rod. Six turns of wire
along a length of 45 mm was for the filament assembly. If thinner Ta wire was
used it can sag and produce an irregular activation temperature and in some cases
damaging the sample surface due to contact. The tantalum wire (Aldrich) used was
99.9 % pure and the elements present are shown in Table 3.1

3.2.2 Deposition onto Micro Drill and Dental Bur Substrate using the VFCVD System

The filament material and filament geometry arrangement were important factors
to consider in order to have improved coatings using the new VFCVD method. To
ensure uniform coating around the dental bur, or micro drill, was positioned centrally and coaxially within the coils of the filament, the six-spiral (coil) filament
was made with 1.5 mm spacing between the coils.

3.3 Experimental Conditions for Diamond Deposition

The CVD reactor consisted of a cylindrical stainless steel chamber, which measured 20 cm in diameter and 30 cm in length. The gas source used during the
deposition process was composed of a mixture containing 1 % methane with an
excess of hydrogen; the volume flow rate for hydrogen was 100 sccm, while the
volume flow rate for methane was 1 sccm. The deposition time and pressure in the
vacuum chamber were 5–15 h and 20 Torr (2.66 kPa), respectively. The substrate
temperature was measured by a K-type thermocouple mounted on a molybdenum
substrate holder. The depositions were carried out between 800 and 1,000 °C. The
filament temperature was measured using an optical pyrometer and found to be
between 1,800 and 2,100 °C depending upon the filament position. A summary

Table 3.2 VFCVD process conditions used for diamond film deposition on dental burs

Process variables	Operating parameters
Tantalum filament diameter (mm)	0.5
Deposition time (h)	5–15
Gas mixture	1 % CH_4 in excess H_2
Gas pressure (Torr)	2.66 kPa (20 Torr)
Substrate temperature (°C)	800–1,000
Filament temperature (°C)	1,800–2,100
DC voltage	HP, 0–40 V, 0–30 Amp
Distance between filament and substrate (mm)	5

of the process conditions optimised to grow good quality polycrystalline diamond films is shown in Table 3.2.

3.4 Substrate Pre-treatments

The characteristics of substrates required for diamond deposition were described in Chap. 1. A high density of crystallites is needed in the early growth stages to form a continuous film. This can be achieved by using a suitable substrate pre-treatment. Silicon wafers used in the semiconductor industry are highly polished and therefore a pre-treatment is particularly important. Several methods of pre-treatment have been reported (XiLing and ZhaoPing 1994; Mitura 1987; Ashfold et al. 1994):

- Scratching the surface with small diamond or SiC μm sized particles.
- Using ultrasonic treatment in slurry of hard grit (e.g. diamond).
- Chemical treatments such as acid etching
- Depositing hydrocarbon thin films on the substrate surface

The basis for most of these methods is to produce scratches or irregularities in the substrate surface. These act as nucleation sites for diamond. Scratching with diamond results in particles being embedded into the surface and assists in the nucleation and subsequent growth of diamond crystallites. However, deposition of diamond onto cemented tungsten carbide (WC-Co) dental burs is problematic due to WC containing cobalt as a binder. Carbon dissolves readily in Co at temperatures used in diamond CVD that causes diamond films to be of poor quality and poorly adhered to the substrate and, at worst, the deposition of amorphous carbon films rather than diamond. There are a number of potential surface treatments, which can be used to overcome these problems including chemical etching, ion implanting, interlayer coating and bias treatment (Ahmed et al. 2000; Hassan et al. 1999; Jones et al. 2003). Most pre-treatments damage the surface, however bias enhanced nucleation (BEN) aids nucleation without substantially damaging the surface of the substrate.

3.4.1 Pre-treatment of Mo/Si Substrates

Prior to pre-treatment Si and Mo substrates were ultrasonically cleaned in acetone for 10 min in order to remove any unwanted residue on the surface. Abrasion with 1 μm diamond powder was performed for 5 min. Alternatively, substrate was immersed in diamond solution containing 1–3 μm of diamond particles and water for 1 h in an ultrasonic bath. These methods produce scratches on the surface, which act as nucleation sites. The substrates were then washed with acetone in the ultrasonic bath for 10 min. The abraded substrate surfaces were characterized by SEM and EDX.

3.4.2 Pre-treatment on WC-Co Dental Bur Substrates

Co cemented tungsten carbide (WC-Co) dental burs, 30 mm in length including the bur head (WC-Co) and shaft (Fe/Cr) and ~1.00 mm in diameter, were used. Prior to pre-treatment substrates were ultrasonically cleaned in acetone for 10 min to remove any loose residues present. Adhesion strength to diamond films is relatively poor, and on cemented carbide surface can lead to catastrophic failure of the coating in metal cutting. The poor adhesion is related to the cobalt binder that is present to increase the toughness of the tool but it suppresses diamond nucleation and causes deterioration of diamond film adhesion. To eliminate this problem, it is usual to pre-treat the WC-Co surface prior to CVD diamond deposition. The following two-step chemical pre-treatment procedure was used. A first step etching, using Murakami's reagent [10 g $K_3Fe(CN)_6$ + 10 g KOH + 100 ml water] was carried out for 10 min. Acetone was used in an ultrasonic bath, followed by a rinse with distilled water (Sein et al. 2002a, b). The second step etching was performed using an acid solution of hydrogen peroxide [3 ml (96 % wt.) and H_2SO_4 + 88 ml (30 % w/v) H_2O_2] for 10 s in order to remove the Co from the surface. The substrates were then washed again with distilled water in an ultrasonic bath. After wet treatment the dental bur was abraded with synthetic diamond powder (1 μm grain size) for 5 min and followed by ultrasonic bath treatment with acetone for 20 min. The etched surfaces of the substrates were characterised by SEM and EDX.

3.4.3 Negative Biased Enhanced Nucleation

By applying a negative bias to the substrate the nucleation density, adhesion and surface properties of the resulting diamond film can be improved (Hassan et al. 1999, Ali et al. 1999). The effects of various process parameters such as bias time, emission current, bias voltage and the filament arrangement on the film properties have been investigated. Nucleation of diamond is an important step in the growth of diamond thin films, because it strongly influences diamond growth, film quality and morphology. Generally, seeding or abrading with diamond powder or immersing in diamond paste containing small crystallites processed in an ultrasonic bath enhances nucleation. One of the most promising in situ method for diamond nucleation enhancement is substrate biasing. In this method, the substrate is biased negative during the initial stage of deposition. Positive substrate biasing can also be applied but is less commonly used.

Plasma consists of a complex soup of positive and negative ions, radicals and neutrals. Negatively biasing the substrates causes positive ions to be accelerated towards the substrate thus enhancing adhesion of the subsequently grown diamond film. In addition, biasing is a highly controlled and non-destructive way of creating nucleation sites for the growth of diamond and ion bombardment also moves atoms around on the surface breaking the diamond clusters into smaller areas.

A negative bias voltage up to -300 V was applied to the substrate relative to the filament. This produced emission currents up to 200 mA. The nucleation times used were between 10 and 40 min. The input gas mixture during BEN was 3 % CH_4 in H_2 at a pressure of 20 torr (26.6 kPa). In the biasing process electrons are emitted from diamond coated molybdenum substrate holder and accelerate towards the filament gaining energy from the electrical field. These energetic electrons enable the ionisation of the hydrocarbon process gases. When the negative bias was applied to the anode the voltage was gradually increased until a stable emission current was established and a luminous purple gas discharge, typical of hydrogen rich atmospheres, was formed near the substrate.

3.5 Machining of Bone with Diamond Coated Tools

The structure of compact bone is heterogeneous, and as such, is difficult to shape by cutting tools during clinical surgical practices. The structure of bone can have a devastating effect on the performance of the cutting tool unless it is coated with a thin diamond film that is wear resistant (Jackson et al. 2007). This section investigates the use of diamond coated cutting tools to prepare bone for dental implants and the implications of their use on the machining characteristics of biological materials.

A micromachining technique is described that easily removes bone without destroying the natural features of the surface of bone. One technique that shows much promise in machining bone is ultra high-speed milling. This technique has been shown to produce micro and nano scale structures in the same way as a conventional machine tool produces macroscale features. A special requirement of machining at such small scales is the need to increase the rotational speed of the cutting tool. The cutting speed of the cutting tool is given by the following equation:

$$V = r\omega \tag{3.1}$$

where V is the cutting velocity (m/s), r is the cutting tool radius (m) and ω is the rotational speed in (radians/s). From this relationship it can be seen that as the cutter diameter reduces in size to create micro and nanoscale features the rotational speed must dramatically increase to compensate for the loss of cutting speed at the micro and nanoscale. At the present time, the fastest spindle commercially available rotates around 460,000 rpm under no-load conditions.

Research is currently underway to improve the performance of air turbine spindles where the initial aim is to reach 1,400,000 rpm (Jackson et al. 2007). Strain rates induced at these high speeds cause chip formation mechanisms to be significantly different than at low speeds. Additionally, it is now possible to experiment at the extreme limits of the fundamental principles of machining at ultra high speeds and at the micro and nanoscales using known theories of machining. This section discusses the use of these theories at the microscale and at high strain rates and discusses the use of a model of initial chip formation during high strain rate deformation at the microscale.

3.5.1 Micromachining

Following the development of modified equations proposed by Shaw (Shaw 1996), the equations will be applied to a 6 flute end milling cutter with a shank of diameter 1.59 mm, cutting diameter 700 μm, rotated at a speed of 250,000 rpm or 26,180 rad/s. The rake angle was $\alpha = 7^0$, clearance angle $\theta = 10^0$ and the shear plane angle $\varphi = 24^0$. The material cut was cancellous bovine femur, and the horizontal force F_h was calculated using the equation:

$$F_h = mr\ \omega^2 \tag{3.2}$$

where m is the tool mass (kg), and r is the tool radius (m). Coefficients of friction between different materials have been investigated by Bowden and Tabor (2001). They also describe methods used for the determination of the coefficient of friction. Using the method of the inclined plane, the coefficient of friction of cancellous bovine femur on tungsten carbide and steel is in the range $\mu = 0.5$–0.6 under lubricating conditions, i.e. sliding on a plane coated with a saline solution. Using the following equation,

$$\beta = \tan^{-1}\mu \tag{3.3}$$

The friction angle β can then be determined under these conditions. It was found to be 31° under experimental conditions. This is in excellent agreement with Merchant and Zlatin's nomograph (Shaw 1996) for calculating the coefficient of friction. The vertical force F_V can be found using the relationship:

$$F_v = \frac{\mu F_h - F_h\ \tan\alpha}{1 + \mu\ \tan\alpha} \tag{3.4}$$

This was found to be 5.25 N. Again, referring to the Merchant and Zlatin's (Shaw 1996) nomograph for the coefficient of friction, the value of F_h can be independently predicted to be 5.33 N. The force perpendicular to the tool plane F is found to be:

$$F = F_h\ \sin\alpha + F_v\ \cos\alpha \tag{3.5}$$

F was determined as 6.67 N. The force normal to the tool plane N is provided using the equation:

$$N = F_h \cos\alpha - F_v \sin\alpha \tag{3.6}$$

where N was found to be 11.1 N. The force perpendicular to the shear plane F_s can now be determined by,

$$F_s = F_h \cos\phi - F_v \sin\phi \tag{3.7}$$

And was estimated to be 8.76 N. The force normal to the shear plane N_s is given by the equation:

$$N_s = F_v \cos\phi + F_h \sin\phi \tag{3.8}$$

where N_s is 9.61 N. Now the frictional force F_f is:

$$F_f = F_v \cos\alpha + F_h \sin\alpha \tag{3.9}$$

F_f is approximately 6.67 N. It is possible to check this value with Merchant and Zlatin's (Doyle et al. 1979) nomograph for frictional force. However, the values for F_h and F_v are so small the extreme limits of the nomograph are being tested so it is difficult to give an accurate value for F_f, it is certain that this value is below 10 N which is in close agreement with the calculated answer. The shear stress τ is found using the following quotient:

$$\tau = \frac{F_s}{A_s} \tag{3.10}$$

which, has a value of 1.8 GN/m². The direct stress ρ is found by applying the relationship,

$$\sigma = \frac{N_s}{A_s} \tag{3.11}$$

σ is found to be 1.95 GN/m². The chip thickness ratio, r, is given by:

$$r = \frac{t}{t_c} \tag{3.12}$$

where t is the un-deformed chip thickness (or depth of cut) and t_c is the measured chip thickness. The machining of bone was conducted at such a small scale that it was difficult to measure t. Therefore r was calculated using the equation:

$$r = \frac{\tan\phi}{\cos\alpha + \sin\alpha \, \tan\phi} \tag{3.13}$$

which yields r = 0.425 and therefore t = 4.25 µm. This is in excellent agreement with Merchant and Zlatin's nomograph for shear angles and the calculation can be made in confidence. Shear strain γ is found from,

$$\gamma = \frac{\cos\alpha}{\sin\phi \cos(\phi - \alpha)} \tag{3.14}$$

γ was found to be 2.55, this can be independently verified from Merchant and Zlatin's nomograph for shear strain which yields a value of 2.51. The cutting velocity V is found using:

$$V = T_{td}\omega \tag{3.15}$$

where, V is 9.1 m/s. The chip velocity is found from applying the following equation:

$$V_c = \frac{\sin \phi}{\cos(\phi - \alpha)} \tag{3.16}$$

where V_c is equal to 3.9 m/s, this can also be found from

$$V_c = rV \tag{3.17}$$

The two results are in agreement with each other. The shear velocity V_s is given by:

$$V_s = \frac{V \cos \alpha}{\cos(\phi - \alpha)} \tag{3.18}$$

where V_s is calculated to be 9.5 m/s. V_s can also be found from:

$$V_s = \gamma V \sin \phi \tag{3.19}$$

Again, the two results are in agreement with each other. The strain rate $\dot{\gamma}$ is given by:

$$\dot{\gamma} = \frac{V \cos \alpha}{\Delta y \cos(\phi - \alpha)} \tag{3.20}$$

where Δy is the shear plane spacing and $\dot{\gamma}$ is found to be 8333×10^3 s^{-1}. The feed rate is 1 mm/min under experimental conditions and the feed per tooth δ is given by,

$$\delta = \frac{F_r}{N\omega} \tag{3.21}$$

where N is the number of teeth. Therefore, Δ is 6.67 μm and the scallop height is found by using the following:

$$h = \frac{\delta}{\left(\frac{4T_{td}}{\delta}\right) + \left(\frac{8N}{\pi}\right)} \tag{3.22}$$

Therefore h is calculated to be 1.59×10^{-11} m under the experimental conditions of machining. These conditions assume that the tool is rotating concentrically about a fixed point, but this is not usually the case with such small tools machining at high speeds. This means that the modelling of chip curl during the initial stages of machining is difficult and one approach is to consider a situation where the bending of the cutting tool produces a multiple number of shear planes rather than a single unified shear plane. This idea is generally suited to machining at the microscale and is considered as a modification of the early assumption made by Ernst and Merchant that a single shear plane model exists during the machining of materials (Shaw 1996). Also, because compact bones are composed of laminates of porous tissue, the orientation of the laminates may indeed control the direction of fracture during the formation of a chip and will inevitably change the direction and angle of the shear plane during machining. Therefore, a simple chip curl model that accounts for the change in laminate direction and tool bending is possible.

3.5.2 Chip Formation

Chip curvature is a significant parameter in machining operations from which a continuous chip is produced. In this paper, observations are made on initial chip curl in the simplified case of orthogonal cutting at the microscale. The cutting process may be modelled using a simple primary shear plane and frictional sliding of the chip along the rake face. When the region of chip and tool interaction at the rake face is treated as a secondary shear zone and the shear zones are analyzed by means of slip-line field theory, it is predicted that the chip will curl. Thus, chip curvature may be interpreted as the consequence of secondary shear. Tight chip curl is usually associated with conditions of good rake face lubrication (Doyle et al. 1979). At the beginning of the cut, a transient tight curl is often observed, the chip radius increasing as the contact area on the rake face grows to an equilibrium value. Thus, it might be suggested that tight curl is an integral part of the primary deformation due to friction interactions between chip and tool edge.

The process of continuous chip formation is not uniquely defined by the boundary conditions in the steady state and that the radius of curl may depend on the build-up of deformation at the beginning of the cut (Doyle et al. 1979). A treatment of primary chip formation at the microscale is presented, which considers chip curl as a series of heterogeneous elements in continuous chip formation at the microscale. The free surface of the chip always displays 'lamellae', which are parallel to the cutting edge. The chip is usually considered to form by a regular series of discrete shear events giving a straight chip made up of small parallel segments. It is assumed that multiple shear planes are created owing to the formation of discrete lamellae. However, no account is taken of how bone material moves passed the tool between shearing events. The following observations follow on from Doyle, Horne, and Tabor's analysis of primary chip formation.

Figure 3.2 shows the instabilities during chip formation (previously defined) that gives rise to instantaneous chip curl. The shaded range of Fig. 3.2b is the consequence of a built-up edge that very quickly becomes part of the segmented chips shown in Fig. 3.2d. This 'material' provides the means to curl the chip and as a consequence of this event, the following model is presented. Previous treatments of chip curl analysis have focused on chip formation with a perfectly stiff cutting tool. However, during the machining of bone it is observed that the cutting tool bends as it cuts. This means that primary chip curl models must account for deflection of the cutting tool by bending during an orthogonal machining operation. Computational approaches to modelling chip formation at the micro-scale have been attempted in recent years by a number of researchers, who have used a molecular dynamics simulation approach using a perfectly stiff cutting tool.

The generation of a transient built-up edge ahead of the cutting tool between shearing events in a bulging-type of motion generates the shape of the segment of the metal chip. This is shown in Fig. 3.2c, with the built-up edge forming the 'shaded triangle' above the shear plane. If it is assumed that the built-up edge does not 'pass' under the tool edge, then the areas of the shaded triangles in Fig. 3.2b, c will be equal. The chip moves away from the rake force in a manner shown in Fig. 3.2d. The radius of chip curl can be calculated by assuming that the built-up edge in transient and that the element of the 'bulged' material contains a small angle relative to the tool and workpiece. This angle will inevitably change during the bending action of the cutting tool. If we assume that the cutting tool moves from point A to point D then the shear plane AC rotates to position HC as the built-up edge from triangle ABD is pushed into the segment of the chip. At point D, the shear along DF begins and segment DHCF is completed. HC and DF meet at R, the centre of the circle of the chip segment. Since the angle HRD is small, RD may be referred to as the radius of the chip. The clearance angle is, θ.

Triangles ABD and HBC are equal in area and the depth of cut FG is equal to d. The spacing between the segments, i.e., the lamellae, is CE, which is equal to BD, which is equal to s. The chip thickness between lamellae, TC, is equal to t, whilst the rake angle SBD is equal to α. The cutting tool bends when machining compact bone, which reduces the effective rake angle to α_b. We know that the chip radius r can be taken as RD, whilst the shear angle subtended is B$\hat{A}$D, or ϕ.

The calculation of the chip radius is provided by the following analysis,

$$DP = s.\cos(\phi - \alpha_b) \tag{3.23}$$

$$AB = \frac{B.S}{\sin \phi} \tag{3.24}$$

where,

$$BS = s.\cos \alpha_b \tag{3.25}$$

Thus,

$$DF = AC = \frac{d}{\sin \phi} \tag{3.26}$$

And,

$$AB = \frac{s.\cos \alpha_b}{\sin \phi} \tag{3.27}$$

Now,

$$BC = AC - AB = \frac{d}{\sin \phi} - \frac{s.\cos \alpha_b}{\sin \phi} \tag{3.28}$$

Fig. 3.2 Instability during the formation of a chip during micromachining: **a** segmented, continuous chip; **b** chip forming instability due to built-up edge; **c** movement of a built-up edge to form a chip; **d** serrated, continuous chip curl. Reproduced with permission. Copyright retained by Inderscience Publishers

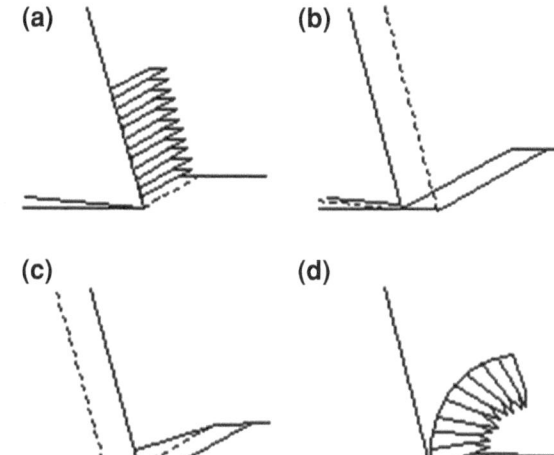

Therefore,

$$BC = \frac{(d - s.\cos \alpha_b)}{\sin \phi} \tag{3.29}$$

The areas of $\triangle ABD$ and $\triangle HBC$ are equal, such that,

$$AB.DP = HQ.BC \tag{3.30}$$

Hence,

$$HQ = \frac{DP.AB}{BC} = \frac{s.\cos \alpha_b.s.\cos(\phi - \alpha_b).\sin \phi}{\sin \phi.(d - s.\cos \alpha_b)} \tag{3.31}$$

And thus,

$$HQ = \frac{s^2.\cos \alpha_b.\cos(\phi - \alpha_b)}{(d - s.\cos \alpha_b)} \tag{3.32}$$

Also,

$$BH = \frac{HQ}{\cos(\phi - \alpha_b)} = \frac{s^2.\cos \alpha_b}{d - s.\cos \alpha_b} \tag{3.33}$$

And,

$$DH = BH + BD = s + \frac{s^2.\cos \alpha_b}{(d - s.\cos \alpha_b)} \tag{3.34}$$

Such that,

$$DH = \frac{s.d}{(d - s.\cos \alpha_b)} \tag{3.35}$$

Hence,

$$CH = \frac{TC}{\cos \theta} = \frac{t}{\cos \theta} \tag{3.36}$$

Therefore,

$$\sin H\hat{R}D = \sin H\hat{C}B = \frac{s^2 \cdot \cos \alpha_b \cdot \cos(\phi - \alpha_b) \cdot \cos \theta}{t \cdot (d - s \cdot \cos \alpha_b)} \tag{3.37}$$

And,

$$\sin D\hat{H}R = \sin T\hat{H}C = \cos \theta \tag{3.38}$$

In triangle HRD,

$$\frac{RD}{\sin D\hat{H}R} = \frac{DH}{\sin H\hat{R}D} \tag{3.39}$$

Therefore,

$$RD = r = DH \cdot \frac{\sin D\hat{H}R}{\sin H\hat{R}D} = \frac{s.d}{(d - s \cdot \cos \alpha_b)} \cdot \frac{\cos \theta.t.(d - s \cdot \cos \alpha_b)}{s^2 \cdot \cos \alpha_b \cdot \cos(\phi - \alpha_b) \cdot \cos \theta} \tag{3.40}$$

Thus,

$$r = \frac{d.t}{s \cdot \cos \alpha_b \cdot \cos(\phi - \alpha_b)} \tag{3.41}$$

If the width of the lamellae, s, is small compared to the chip thickness, then for continuous machining with a single shear plane,

$$\frac{d}{t} = \frac{\sin \phi}{\cos(\phi - \alpha_b)} \tag{3.42}$$

Hence,

$$\frac{t}{\cos(\phi - \alpha_b)} = \frac{d}{\sin \phi} \tag{3.43}$$

And so,

$$r = \frac{d^2}{s \cdot \cos \alpha_b \cdot \sin \phi} \tag{3.44}$$

Equation 3.44 predicts that a positive chip radius will occur at negative rake angles. The approximations considered in this model are appropriate when one considers that the model assumes that a secondary shear plane exists and that multiple primary shear planes exist at discrete intervals of time.

3.5.3 Experimental

3.5.3.1 Micromachining Apparatus

The machining of bovine femur was performed using a modified machining centre. The micromachining centre was constructed to incorporate a high-speed air turbine spindle rated to operate at 460,000 rpm under no load conditions. When operating at relatively deep depths of cut, the speed of the spindle decreases to approximately 320,000 rpm. The table of the machine tool was configured to move in x-y-z co-ordinates by attaching a cross-slide powered by a d.c. motor, in all three principal axes. Each motor was controlled by a Motionmaster^{TM} controller with a resolution as low as 500 nm. The cutting tools used were coated with diamond. The bovine femur samples were machined at various depths of cut at high speed and were machined in an aqueous saline solution. The cutting tools were inspected at the end of all machining experiments using an Environmental Scanning Electron Microscope. The measured spindle speed was 320,000 rpm during the machining experiments. The depth of cut ranged between 50 and 100 μm for all machining experiments. The machining feed rate was conducted at 5 mm/s (0.3 m/min). The microscale cutting tool used was 700 μm in diameter (microscale) and was associated with a cutting speed of 117 m/min and a machining feed rate of 0.3 m/min. The results of the experimental procedures are shown in Table 3.3. The machined chips were examined in an environmental scanning electron microscope where the lamellar spacing on each chip was determined. Transient chip curl was measured at the first 90° of tight chip curl. The curl radii were compared with the calculated value derived using the idealized model, taking into account the degree of bending of the cutting tool.

3.5.3.2 Observations of Machined Chips

There are significant differences in the size and shape of chips when machined at medium and high speeds. This is especially so for biological materials such as compact bone. Figure 3.3 shows a collection of chips machined from bovine femur. It is seen in Fig. 3.3 that many of the particles are in fact chunks of material rather than nicely formed chips. It is possible that the chunks were formerly parts of larger chips that have since broken down and that chip thickness values should be recalculated based on the larger chip size.

It can also be seen that the chips in Fig. 3.3 are more consistent in terms of length, width, and depth. Their lamellar spacing is also regular in period, which would indicate that cutting conditions at high speed are stable.

Single chip formations are shown in Figs. 3.4 and 3.5. While the width observed is similar to that for low speed cutting, the chip length of high-speed chips is much shorter than low speed chips. This could be because at low speed the chip has a greater time in contact with bone thereby removing more material, which is reflected in the increased chip length. One of the major differences observed between low and high speed micromachining of bone is in the spacing of

Table 3.3 Experimental data comparing initial chip curl during micromachining and initial chip curl predicted by the model

Rake angle after bending (°)	Shear plane angle (°)	Mean lamellar spacing (μm)	Observed chip curl (mm)	Calculated chip curl (mm)
22	37	0.98	17.55	18.01
15	25	1.55	14.42	14.65
8	18	1.9	16.55	17.1
3	12	2.95	15.82	16.22

The depth of cut was 100 μm. Note that variable primary shear plane is observed during the experiments. Reproduced with permission. Copyright retained by Inderscience Publishers

Fig. 3.3 Characteristic chip shapes cutting bovine femur at high speeds. Reproduced with permission. Copyright retained by Inderscience Publishers

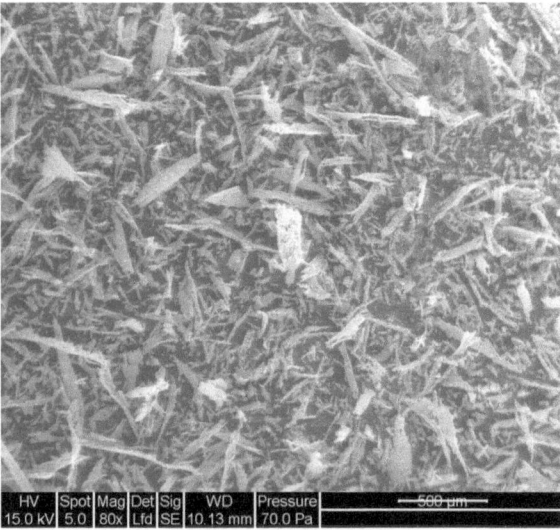

Fig. 3.4 Individual chip formation at high speed. Reproduced with permission. Copyright retained by Inderscience Publishers

Fig. 3.5 Lamellae spacing of bovine femur at high speed. Reproduced with permission. Copyright retained by Inderscience Publishers

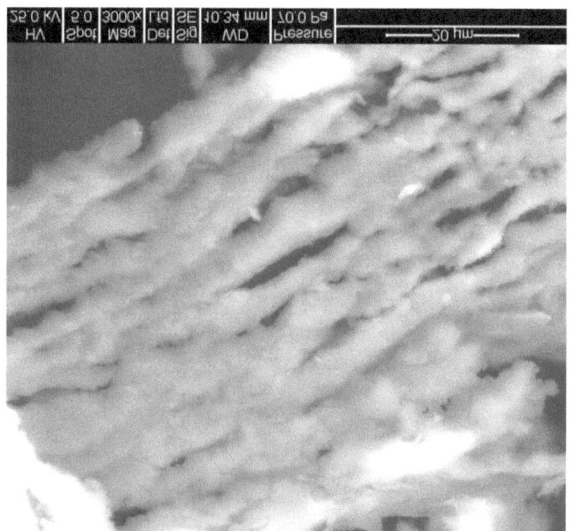

Fig. 3.6 Magnified image of the cutting tool showing cutting edges and adhered bone material. Reproduced with permission. Copyright retained by Inderscience Publishers

the lamellae. In low speed cutting, the chip spacing varies by a significant amount. However, at high cutting speeds the spacing is regular in period. At high speeds this process is accelerated to an extremely high level as the strain rate calculations have shown. In fact experiments show that chip types are similar in other materials such as metals.

Figure 3.6 shows a magnified image of a coated cutting tool. The clearance faces of the flutes of the cutting tool show adherent bone chips with finely striated

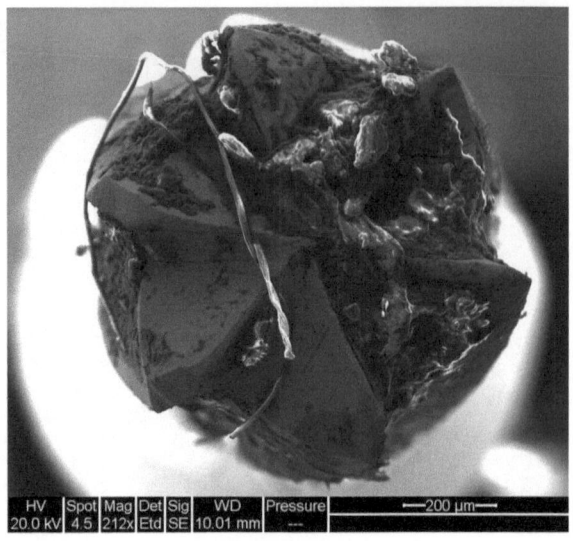

lamellae, as noted on the left hand side of the tool. Figure 3.7 shows a magnified image of a coated cutting tool detailing the cutting edge and its relationship to the adherent film of bovine femur showing fine striations of lamellae generated at high strain rates.

3.5.3.3 Micromachining Results

The results of machining bovine femur at the microscale are compared to the model described for primary chip curl during the primary stages of chip formation. It should be noted that all results presented in Table 3.3 are for bone machined in an aqueous saline environment. Table 3.3 shows the results for biomachining using a variety of rake angles. It should be noted that bending of the cutting tool produces a less acute rake angle when machining takes place. However, the shear plane angle is increased and larger chips are produced.

3.6 Discussion

It can be seen from the above analysis that despite the extremely high strain rates imposed due to high speed cutting, macroscale equations can be applied accurately and produce impressive results. The most significant differences however appear in the following categories: strain rate, scallop height, and chip type. Many of the forces are of a similar order of magnitude offering no significant difference between macro low speed and micro high speed machining. This is important during tool design as small tools must absorb the same impact forces as larger tools do during impact. However, when considering the strain rate it can be seen during micro high speed machining the strain rate is $8{,}333 \times 10^3$ s^{-1} compared to the macro low speed case of 667×10^3 s^{-1}, a 12.5 times increase which relates directly to a 12.5 increase in speed from 20,000 to 320,000 rpm. The increase in strain rate is directly related to the increase in cutting speed, this is expected as the cutter is imparting the strain and therefore a rate of strain to the material. The lamellae spacing Δy in Eq. 3.20 has a significant effect on the strain rate, comparing macro and micro scale chips it is found that lamellae are ten times more closely packed in the high speed chips than the low speed chips. The purpose of machining is to create surfaces that are useful, hence surface quality should be an important consideration of milling, a measure of this is scallop height.

An improvement is seen in the micro high-speed case with a scallop height of 1.58×10^{-11} m compared to 8.9×10^{-9} m for macro slow speeds. Although both values seem insignificant it must be remembered at the micro and nano scales post process finishing is inappropriate, therefore created structures must be produced to specification without further processing. Additionally, owing to the aspect ratio small imperfections become serious defects at small scales. From the calculations it can be seen that there is an improvement in the scallop height, which is not the improvement required when considering the scale order of magnitude has changed by a factor of four. This is because the current spindle speeds reached are not high enough for effective machining, if this speed is increased to 1,400,000 rpm then the orders of magnitude are increased further still.

The experimental results and observations provide an interesting view of machining bone at the microscale. When one considers the approximations made in the derivation of the chip curl model, the experimentally measured results compare well with the calculated chip curl. This indicates that cutting tool bending contributes significantly to initial chip curl prior to any significant frictional interactions on the rake face of the cutting tool. The proposed model describes the initial stages of chip curl quite well. If the description of chip curl is accurate, then continuous chip formation at the microscale needs to be re-investigated. If one considers the movement of the cutting tool, from point A towards point D, we expect the shear plane to oscillate between AC and HC depending on the amount of energy required to move the built-up edge into the segment of the subsequent chip. The cycle begins again when accumulated material is deposited on to the edge of the cutting tool then on to the subsequent segment of the chip produced during machining.

3.7 Conclusions

The equations of metal cutting can be applied in the high-speed microscale environment. The nomographs of Merchant and Zlatin can be applied confirming that future calculations can be compared to these well-constructed charts. High strain rates change the mechanism of chip formation thereby altering the shape of the chip. Also, high strain rates appear to provide less dependence on material properties in determining chip formation and shape.

A model of chip curl at the microscale has been developed and agrees well with experimental data. It appears that the bending of the cutting tool contributes significantly to the primary chip prior to significant frictional interactions on the rake face of the cutting tool. It is shown that primary chip curl is initiated by the amount of material deposited onto the cutting tool that manifests itself as a wedge angle that controls the amount of material pushed into the base of the segment of the chip between oscillations of the primary shear plane. The future development of this technique lies in the ability to rotate cutting tools at extremely high spindle speeds.

The experimental results also suggest that a number of primary shear planes are created during the initial stages of chip formation that contradicts the assumptions made by Ernst and Merchant. Although their experiments were characterized by machining soft, ductile metals at low speeds, it seems appropriate to suggest that their model cannot be initially applied to the machining of laminate structures such as bone. However, a series of single shear planes dominated by dynamic shearing events may describe the machining of a laminate structure. Further investigations on the primary causes of shearing in bone and how they are modelled are required, in addition to how diamond coatings may affect shearing and reduce heat at the contact zone.

Acknowledgments The authors are grateful to Inderscience for allowing the authors to reproduce material published in the International Journal of Nano and Biomaterials, 2009, volume 2, number 6, p. 505. Inderscience retains copyright of the material used in this chapter.

References

Ahmed W et al (2000) CVD diamond: controlling structure and morphology. Vacuum 56(3):153–158

Ali N et al (1999) Role of surface pre-treatment in the CVD of diamond films on copper. Thin Solid Films 355:162–166

Ashfold MNR et al (1994) Thin film diamond by chemical vapour deposition methods. Chem Soc Rev 23(1):21–30

Bowden FP, Tabor D (2001) The friction & lubrication of solids. Oxford Science Publications, Clarendon Press, Oxford, pp 73–75, 83–85

Doyle ED, Horne JG, Tabor D (1979) Frictional interactions between chip and rake face in continuous chip formation. Proc R Soc London A366:173–183

Hassan IU et al (1999) An investigation of the structural properties of diamond films deposited by pulsed bias enhanced hot filament CVD. Thin Solid Films 355:134–138

Jackson MJ et al (2007) Design and manufacture of high speed spindles for dry micromachining applications. Int J Nanomanufacturing 1(5):641–656

Jones AN et al (2003) The impact of inert gases on the structure, properties and growth of nanocrystalline diamond. J Phys Condens Matter 15(39):S2969

Mitura SW (1987) Nucleation of diamond powder particles in an RF methane plasma. J Cryst Growth 80(2):417–424

Sein H et al (2002a) Application of diamond coatings onto small dental tools. Diam Relat Mater 11(3):731–735

Sein H et al (2002b) Application of diamond coatings onto small dental tools. Diam Relat Mater 11(3–6):731–735

Shaw MC (1996) Metal cutting principles. Oxford Science Publications—Series on Advanced Manufacturing, Clarendon Press, Oxford, pp 18–46

Wang WL et al (2000) Nucleation and growth of diamond films on aluminum nitride by hot filament chemical vapor deposition. Diam Relat Mater 9(9–10):1660–1663

XiLing P, ZhaoPing G (1994) Morphologies and adhesion strength of diamond films deposited on WC—6 % Co cemented carbides with different surface characteristics. Thin Solid Films 239(1):47–50

Chapter 4
Diamond Deposition onto Flat Substrates

Abstract CVD of a range of thin films has been employed extensively in the semiconductor industry where, in addition to the required film properties, the reliability and uniformity are key requirements. Diamond deposited onto silicon substrates has been widely researched using both microwave and hot filament chemical vapour deposition (CVD). In this chapter hot-filament CVD has been employed to deposit diamond onto flat substrates to prove that the process is reliable and reproducible in terms of crystallinity and thickness and to study the adhesion onto the substrates used. It is problematic to deposited diamond onto metallic substrates such as stainless steel and therefore interlayers of TiN or TiC were employed to provide improved adhesion. The characteristics of the interlayers and the diamond films have been determined.

Keywords Chemical vapour deposition · CVD · Hot filament · Flat substrates · Interlayers · Titanium nitride · Surface characteristics · Structure and morphology

4.1 Introduction

Considerable research has been carried out on CVD diamond deposition onto flat substrates particularly silicon substrates using a flat horizontal filament arrangement (Ashfold et al. 1994; May et al. 1995; Afzal et al. 1998; Ahmed et al. 2000; Eccles et al. 1999). The mechanisms of growth have also been investigated (Foord 2001). The diamond have also been doped with various elements such as boron and phosphorus to alter its electrical properties (Obraztsov et al. 1998). Diamond films deposited show good adhesion onto silicon substrates using this method provided appropriate substrate treatment is used such as manual abrasion and substrate biasing. Therefore, it is pertinent to prove that the new VFCVD system with horizontal filament arrangement employed in this study used in a conventional arrangement can grow good quality diamond on silicon substrates so that a comparison can be made with the work of other researchers who have used the HFCVD.

For good adhesion the substrate material must form a carbide layer as discussed in Chap. 1. When diamond is deposited on non-diamond substrates an interfacial carbide layer formation is necessary for diamond to grow and adhere successfully. It also partially relieves stresses occurring at the interface. These stresses are thought to be due to lattice mismatch or differences in the thermal expansion coefficients between the two materials (May 1995). If the mismatch is too large then the films deposited tend to crack and eventually peel off. However, when the thermal expansion coefficients are similar then good adhesion is achieved if the substrate readily forms a carbide layer.

4.2 Deposition of Diamond onto Flat Substrates by VFCVD System with Horizontal Filament Arrangement

In this section deposition of polycrystalline diamond onto flat substrates has been studied, particularly silicon after a variety substrate treatments. Silicon has been used because it is the most widely used and extensively investigated substrate material for diamond growth using CVD. Direct deposition of CVD diamond onto Si without pretreating the substrate is problematic due to poor nucleation, which prevents film formation. As mentioned in Chap. 3, for the growth of a continuous film, a sufficient crystallite density must be present after initial growth phase. Silicon wafers used in the electronics industry are very smooth due to polishing processes employed for the preparation to for integrated circuits. For such applications the single crystal wafers must be free of defects. However, for diamond film growth these wafers require abrasion to initiate growth. The scratches on the surface due to abrasion act as nucleation sites for subsequent diamond growth and the formation of a continuous film. Therefore, a variety of methods have been investigated and used for substrate abrasion and nucleation:

- Mechanical abrasion with ~μm sized small hard grit (e.g. diamond, silicon carbide)
- Ultrasonication of samples in slurry of hard grit (e.g. diamond).
- Bias enhanced nucleation (BEN) (Negative/Positive substrate biasing)

The first two of these methods scratch the surface and produce many defects and nucleation sites. The diamond grit may also embed into the surface onto which nucleation can occur readily. Mechanical abrasion is less controlled than ultrasonification and substrate biasing but is still highly effective for diamond growth. However, for applications where a more precise creation of nucleation sites is required the latter two methods are preferred (Hassan et al. 1999).

4.2.1 Experimental Conditions Relating to Flat Substrates

All the silicon substrates (5 × 5 mm) were washed with acetone solution in an ultrasonic bath for 10 min to remove any residues left on the surface. For the Si substrates that were manually abraded 1 μm diamond powder was used for 5 min. The abraded Si

Fig. 4.1 Diamond growths onto abraded silicon substrate for 30 min (1 % CH_4 in 99 % H_2) using VFCVD system in conventional mode

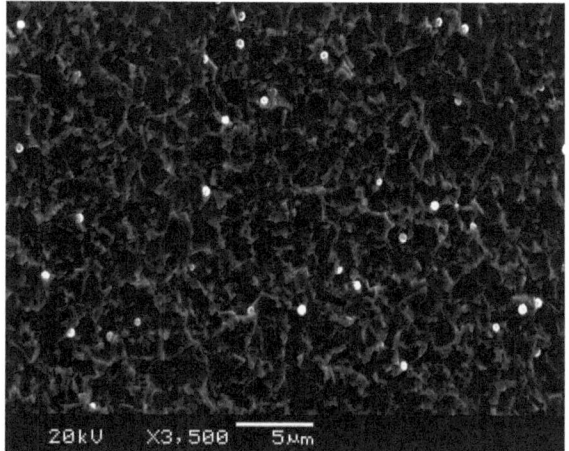

was then rinsed off with the distilled water in an ultrasonic bath. For the bias enhanced nucleation (BEN) study the cleaned Si was placed onto the substrate holder in the CVD chamber and a bias of -200 V was applied to the substrate relative to the filament for 10 min. The input gas mixture was 3 % CH_4 in H_2. During the CVD growth process in a hot filament reactor a plasma is created around the filament and substrate region. The plasma generated in the reactors is composed of a complex "soup" of positive and negative ions, molecules, atoms and radicals. When the substrate is biased negatively the potential difference between the positive ions in the plasma and the negatively charged substrate causes the positive ions to accelerate towards the substrate. The resulting ion bombardment is at a sufficiently high energy to strike the surface creating disruption to the substrate surface creating surface damage, which act as nucleation sites for diamond deposition. Finally, CVD diamond growth was carried out on the pre-treated substrates. The CVD process conditions use have been described and specified in Chap. 3.

4.2.2 CVD Diamond onto Silicon Substrates

The SEM micrograph (see Fig. 4.1) shows that nucleation sites have been created prior to diamond deposition by using mechanical abrasion with diamond powder. After a deposition time of 1 h some diamond crystallites started to nucleate on the abraded substrate. The SEM micrograph shows the transition between the nucleation stage and columnar diamond growth to form the polycrystalline film.

For diamond deposition of 5 h the resulting diamond film morphology and growth are shown in Fig. 4.2.

Evidently Fig. 4.2a shows that diamond deposition on the manually abraded Si for 2 h shows small crystals of diamond started to grow onto the Si substrate. However, after 5 h of deposition uniform diamond films with distinct crystallites can be seen displaying a rough surface (Fig. 4.2b. The crystals are hexagonal shape with crystallite size being of the order of 3–5 μm. This type of film morphology has been widely reported

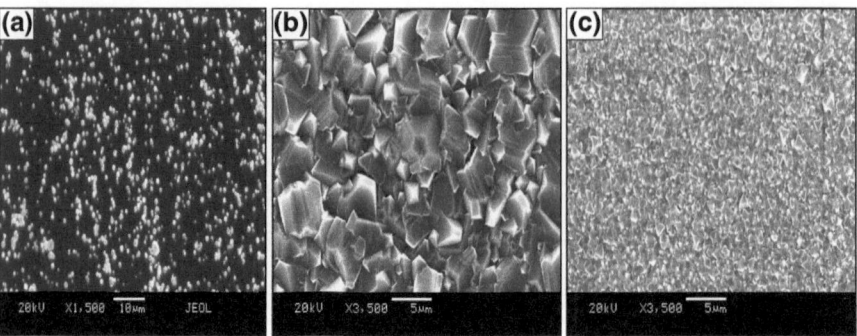

Fig. 4.2 VFCVD systems with horizontal filament arrangement (operated in a conventional mode) diamond films on silicon substrates **a** after a 1-h deposition time **b** after 5-h deposition time **c** 200 V negative substrate bias after 5-h deposition

in the literature. This result gave confidence that our VFCVD system when operating in a conventional way was performing in a similar manner to those previously reported in the literature. When a -200 V was applied to the substrate the diamond crystal morphology appear to alter significantly (Fig. 4.2c). The crystal sizes were much smaller than the unbiased deposition as expected. This is due to the additional process of positive ion bombardment of the nucleation and growing film under biased conditions. The ion bombardment enhances the surface diffusion and breaks up larger nuclei into smaller ones during the initial nucleation stages.

4.3 CVD Diamond onto TiN Interlayer

As explained in Chap. 1 ferrous metals, such as Co, Ni, Fe and steels, were problematic substrates for diamond growth because of their high carbon solubility. A good way to overcome this problem is to create a suitable barrier layer such as TiN or TiC to reduce carbon diffusion into to the metal substrates. Such interfacial layers can be deposited using various techniques including thermal and electron beam evaporation or RF sputtering and magnetron sputtering or ion-assisted deposition. The key purposes of the interlayer are to prevent carbon diffusion and to reduce mismatch between the thermal expansion coefficient between the diamond and the substrate material. In this section the results of our study on the growth of diamond film onto TiN as an interlayer is described. This work was carried out in order to determine whether TiN is a suitable interlayer for subsequent diamond growth.

4.3.1 Experimental Related to the Deposition of the TiN Interlayer

In this study DC reactive Magnetron sputtering was used to deposit TiN films onto the manually abraded Si samples; the TiN film was deposited in an argon

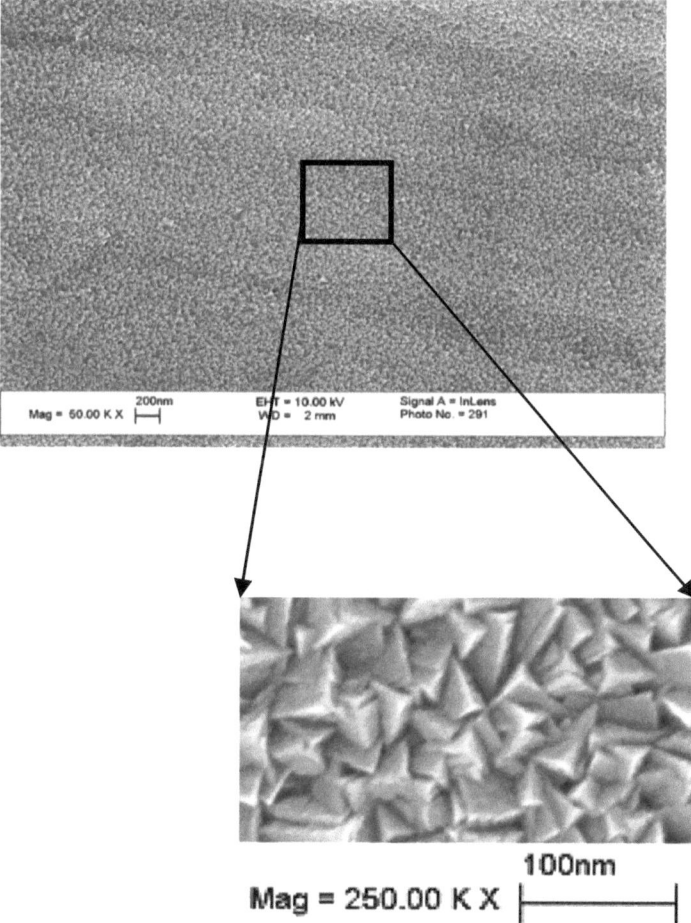

Fig. 4.3 SEMs of TiN interlayer on silicon by DC reactive magnetron sputtering followed by diamond growth using CVD

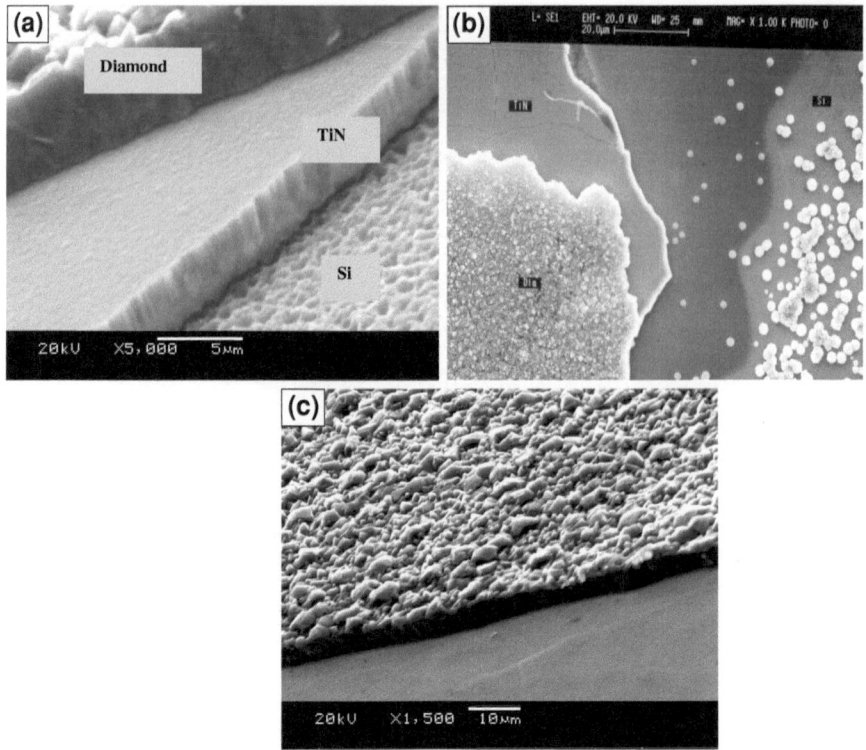

Fig. 4.4 VFCVD systems with horizontal filament arrangement operated in conventional mode showing diamond growths onto silicon substrate with TiN interlayer **a** cross section, **b** top view and **c** oblique of diamond on TiN

atmosphere also containing nitrogen as a reactive gas for 1 h producing films of thickness of 2–3 μm. The purity of the gases was 99.999 % and the target material was pure titanium.

In order to reduce the grain size and simultaneously improve adhesion of the diamond onto the substrate negatively bias enhanced nucleation was employed. Manual abrasion of the TiN interlayer on Si was not suitable in this case because the delicate interlayer would be completely destroyed with this substrate preparation technique.

4.3.2 Diamond Deposition onto TiN Interlayer

It is clear from Fig. 4.3 that the TiN grown as an interlayer has a rough surface on a nanometer scale, which would be highly suitable for diamond growth because it would provide sufficient nucleation sites.

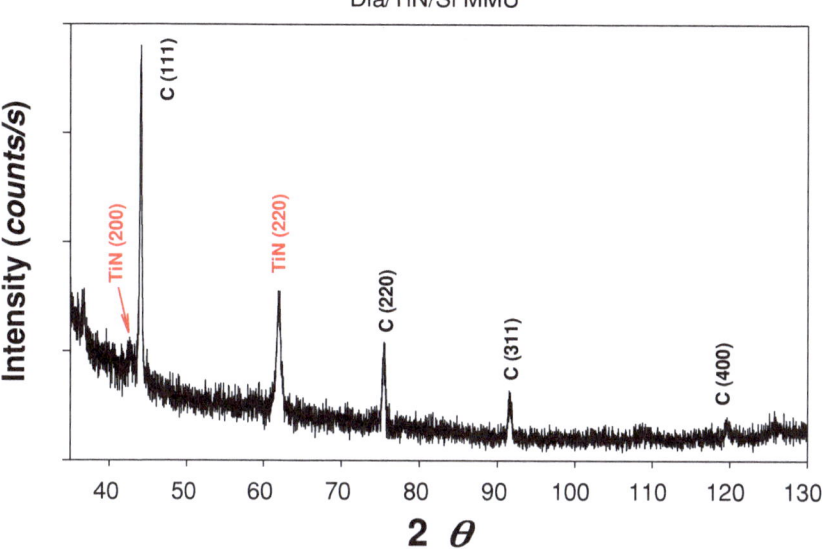

Fig. 4.5 XRD of conventional mode VFCVD system with horizontal filament arrangement diamond films onto silicon substrates with TiN interlayer

As mentioned earlier the delicate TiN interlayer could not be manually abraded. Therefore, substrate biasing was used where a negative bias voltage from −140 to −300 V was applied to the substrate relative to the filament and the methane concentration was kept at 3 %. This produced emission currents of up to 200 mA. The nucleation time used was approximate 30 min.

Figure 4.4a shows a cross section of the VFCVD diamond deposited onto the TiN layer on the silicon using a conventional arrangement. Negative biasing (−200 V) of the substrate relative to the filament with methane concentration at 3 % was carried out for 30 min. After the nucleation period the CH_4 concentration was reduced to 1 % and standard diamond deposition conditions were used for 5 h. It is evident that the TiN layer is much smoother than the diamond film and is of the order of about 2 micrometer as determined from the SEM. The diamond film has a thickness value of 5 micrometer. The diamond deposition time for 5 h give a deposition rate of 1 μm/h, whereas the deposition time for TiN deposited using magnetron sputtering was for 1 h giving a much higher growth rate. The top view Fig. 4.4b shows cracks in the TiN films due to the thermal mismatch between the three materials diamond, TiN and silicon. Figure 4.4c shows the characteristic rough surface generally observed with polycrystalline diamond films.

The XRD spectra (Fig. 4.5) of diamond films deposited onto the TiN interlayer showed a greater diamond formation on the samples, which had been negatively biased prior to continuous growth. In addition, diamond peaks associated with (111) and (220) were evident for the bias assisted sample after diamond VFCVD

Fig. 4.6 Typical Raman
spectrum of a conventional
mode VFCVD diamond film
grown onto silicon substrate
with a TiN interlayer

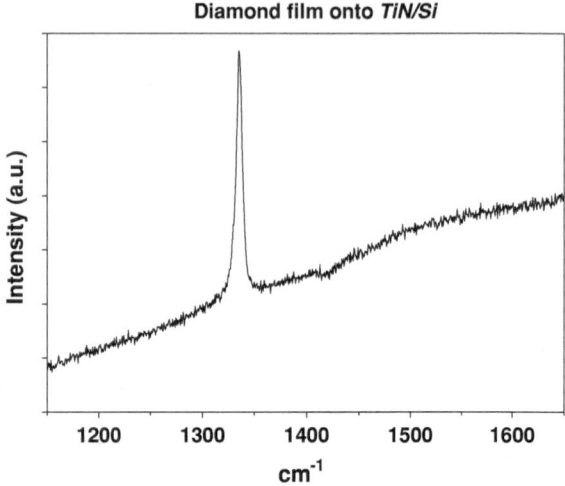

growth. The TiN films deposited on the pre-treated silicon wafers at room temperature exhibited a slightly visible (200) reflection. For TiN this should be the most intense XRD peak. TiN particle size was determined from the FWHM of (200) and (220) TiN peaks and was found to be ~50 nm.

Raman spectroscopy is a standard technique for determining whether the film is indeed diamond as opposed to diamond like carbon (DLC) or graphite. The Raman analysis of the as-grown diamond films (Fig. 4.6) shows that a good quality diamond film has been deposited on the interlayer TiN as is evident from a peak at 1,333 cm^{-1}. As mentioned earlier a sharp peak at 1,332 cm^{-1} indicates that a good quality crystalline diamond film was formed. There was no evident of peaks arising from graphitic growth.

4.4 Conclusions

It is evident that using VFCVD system developed for this study when used in a conventional mode (applied with horizontal filament arrangement) diamond films grown were of a very high quality as evidenced from Raman and SEM analysis. The characteristics of the diamond films were similar to those reported in the literature such as a diamond peak at around 1,332 cm^{-1} and the film being composed of polycrystalline diamond crystals. In addition diamond was also grown on the TiN interlayer using substrate biasing instead of manual abrasion. The promising results presented in this chapter enabled a smooth transition from flat substrates to complex 3-D substrates used in dental burs and drills to be readily achieved.

References

Afzal A et al (1998) HFCVD diamond grown with added nitrogen: film characterization and gas-phase composition studies. Diam Relat Mater 7(7):1033–1038

Ahmed W et al (2000) CVD diamond: controlling structure and morphology. Vacuum 56(3):153–158

Ashfold MNR et al (1994) Thin film diamond by chemical vapour deposition methods. Chem Soc Rev 23(1):21–30

Eccles AJ et al (1999) Influence of B- and N-doping levels on the quality and morphology of CVD diamond. Thin Solid Films 343–344:627–631

Foord JS et al (2001) Reactions of xenon difluoride and atomic hydrogen at chemical vapour deposited diamond surfaces. Surf Sci 488(3):335–345

Hassan IU et al (1999) An investigation of the structural properties of diamond films deposited by pulsed bias enhanced hot filament CVD. Thin Solid Films 355:134–138

May P (1995) Synthetic diamond: emerging CVD science and technology: edited by Harl E. Spear and John P. Dismuhes. Wiley, Chichester, pp 663 1994. ISBN 0 4715 3589 3. Endeavour 19(1):48

May PW et al (1995) CVD diamond-coated fibres. Diam Relat Mater 4(5–6):794–797

Obraztsov AN et al (1998) Direct measurement of CVD diamond film thermal conductivity by using photoacoustics. Diam Relat Mater 7(10):1513–1518

Chapter 5
Diamond Deposition onto Wires and Microdrills Using VFCVD

Abstract Deposition of uniform and adherent diamond onto 3-D substrates is highly challenging particularly when using a conventional horizontal hot filament chemical vapour deposition (HFCVD). In this chapter the deposition of diamond onto wires and microdrills using a new VFCVD system has been described. In this arrangement the wire and drills were held concentrically within the core of the filament where maximum plasma intensity is prevalent. The structure and morphology studies have shown that uniform adherent diamond films have beed deposited using VFCVD. These films were tested for performance and life and the arrangements used and results obtained have been presented.

Keywords Diamond · 3-D substrates · Wires · Microdrills · Structure · Morphology · Performance testing.

5.1 Introduction

The most common methods employed to grow polycrystalline diamond films are hot filament CVD and microwave CVD. However, the vast proportion of literature has been focused on studies involving flat samples, most commonly on silicon. May et al. (1994, 1995, 1994); May 1995) demonstrated that thick adherent diamond films can be deposited onto both metallic and non-metallic wires and fibres (Figs. 5.1 and 5.2). They used a horizontal filament arrangement with the substrate placed outside the filament coils to deposit the thick coatings. However, in this work a VFCVD system with the substrate placed within the coils has been employed for the first time.

Initial studies concentrated on the deposition onto molybdenum (Mo) wires to demonstrate the applicability of diamond growth using a vertical filament arrangement followed by work on diamond CVD onto cemented tungsten carbide microdrills.

W. Ahmed et al., *Chemical Vapour Deposition of Diamond for Dental Tools and Burs*, SpringerBriefs in Materials, DOI: 10.1007/978-3-319-00648-2_5, © The Author(s) 2014

Fig. 5.1 Diamond coated
tungsten wire (Ashfold et al.
1994; May et al. 1994, 1995;
May 1995)

Fig. 5.2 Diamond coated
Tyranno fibre (Ashfold et al.
1994; May et al. 1994, 1995;
May 1995)

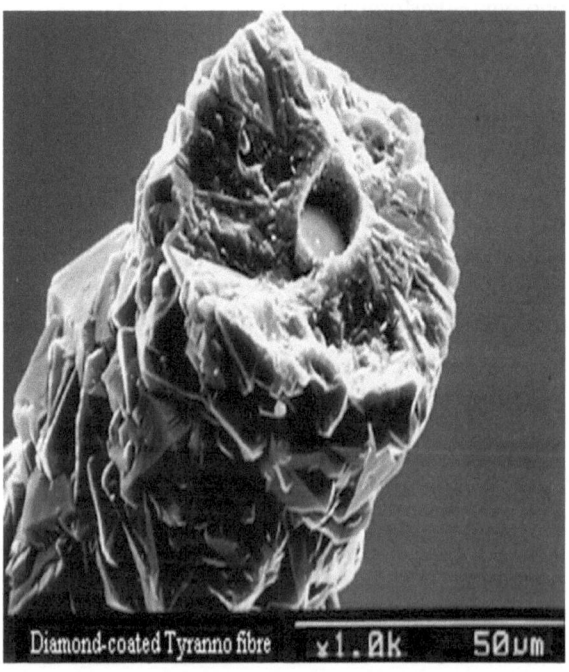

Fig. 5.3 Untreated surface
of molybdenum wire

It is challenging to grow CVD diamond films onto cutting tools with a complex 3-D geometry (Lu et al. 1995; Sein et al. 2002, 2004a). Wires have a cylindrical shape and can be used as a model system for optimising the deposition of diamond on cutting tools such as microdrills and dental burs.

5.2 Diamond Deposition on Molybdenum Wire Using VFCVD

Mo wire with a 30 mm length and 0.5–1.0 mm diameter were used as the substrates. The wire was ultrasonically cleaned in acetone for 10 min to remove loose residues from the surface prior to pre-treatment. Figure 5.3 shows an SEM of the untreated Mo wire surface. It is apparent that the surface is non-uniform on the micron scale and linear striations are clearly visible. Theses striations are formed when the Mo is drawn through a die during the wire manufacturing process.

As explained in Chap. 3 the nucleation stage is an important step in the growth process, which strongly influences the diamond growth, film quality and morphology Zhang and Buck (2002, 2000). In order to enhance nucleation on the substrate the Mo wire was placed in the ultrasonic bath, with diamond powder, of 0.1–1 µm sizes, slurry solution for 30 min. The substrates were then washed again with distilled water in an ultrasonic bath. The treated surfaces of the Mo substrates were characterised by SEM. Figure 5.4 shows that some diamond powder is embedded in the Mo and acted as nucleation centres for diamond growth.

The Mo wires were deposited with VFCVD diamond using the modified vertical filament reactor described in Chap. 3. Before VFCVD diamond deposition the filament was pre-carburised for 30 min with 3 % methane in excess hydrogen at 20 Torr in order to reduce the tantalum evaporation during diamond deposition (Wang et al. 2001; May et al. 2006). After filament carburisation VFCVD was

Fig. 5.4 Molybdenum
surface treated with diamond
slurry

Fig. 5.5 VFCVD diamond
on treated Mo substrate after
2 h depositions

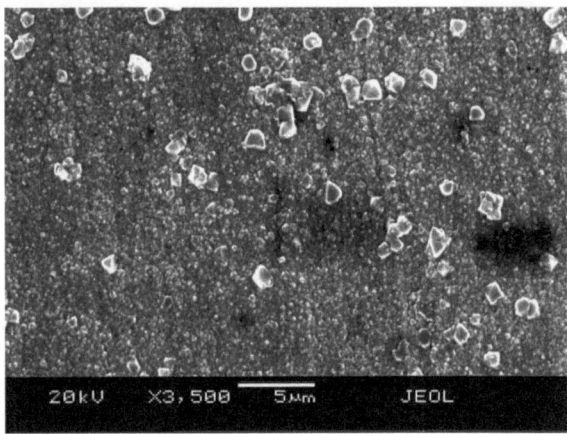

carried out using standard conditions (1 % CH_4 in H_2) for various times up to 5 h.
Film thickness and morphology were determined by using a SEM and film quality
was gauged by Raman spectroscopy.

Previously VFCVD deposition was carried out for 2 h in order to study the
growing diamond prior to complete film coverage. The SEM image in Fig. 5.5
shows that small grains of diamond are deposited on the pre-treated substrate
(Fig. 5.5).

After a deposition time of 5 h a continuous film of polycrystalline diamond was
obtained (Fig. 5.6). The film morphology has good radial uniformity and was composed of highly crystalline diamond. However, the diamond film deposited outside
the filament zone is neither continuous nor uniform with poor diamond coverage
from some regions (Fig. 5.6). This non-uniformity is believed to be due to substrate temperature variations. Outside the filament zone the temperature is lower

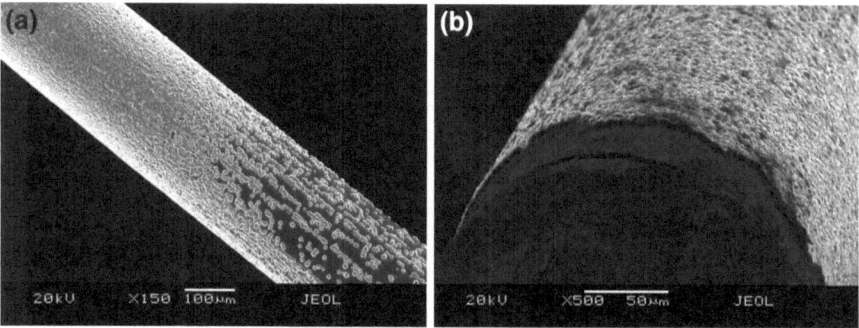

Fig. 5.6 VFCVD diamond on molybdenum rod. **a** Lateral distribution and **b** radial uniformity

Fig. 5.7 Cross sectional
SEM of diamond deposited
using VFCVD on
molybdenum wire

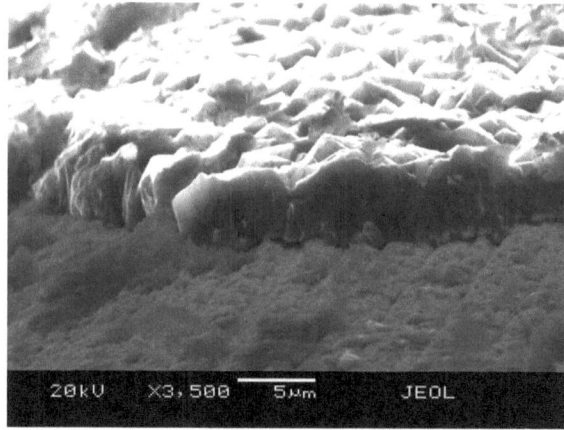

resulting in a lower deposition rate. It is also evident that diamond preferentially nucleates on striations on the Mo wire surface.

Figure 5.7 shows a cross section of the CVD diamond film on molybdenum wire at the centre of the filament where a complete film is formed. The SEM shows the film thickness around the wire to be uniform with an average film thickness of about 5 μm. The diamond film growth rate is therefore 1 μm per hour, which is typical for HFCVD. The film has a columnar structure with a rough surface. Hence this type of film structure is suitable for grinding type applications and is typical of diamond films grown under standard deposition conditions.

Figure 5.8 shows the detailed structure of the VFCVD diamond film. It is evident that the film is polycrystalline and uniform around the wire. The crystallites are variable in size ranging from 2 to 7 μm with an average crystallite of about 4 μm. <111> faceted octahedral grains predominate.

To confirm the quality of the film Raman analysis was carried out. The Raman spectrum (Fig. 5.9) shows an intense sharp peak at 1,332.6 cm^{-1} which is the fingerprint for good quality sp^3 diamond. This result is similar to those reported in the literature.

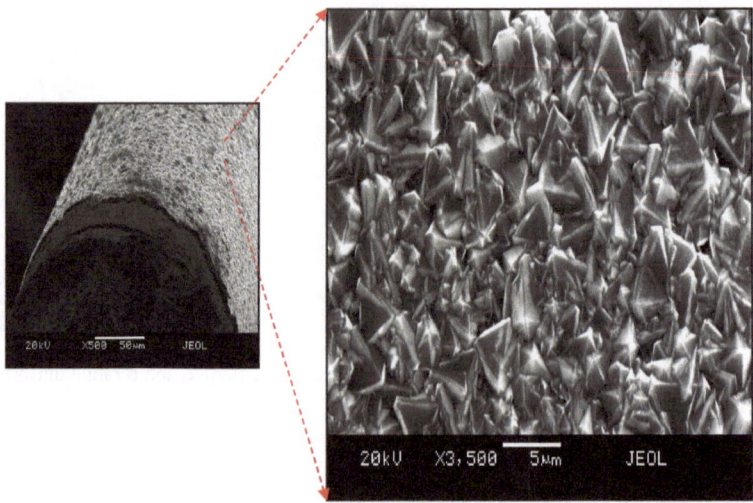

Fig. 5.8 Diamond film on Mo wire. Uniform growth of (111) faceted octahedral diamond film on Mo

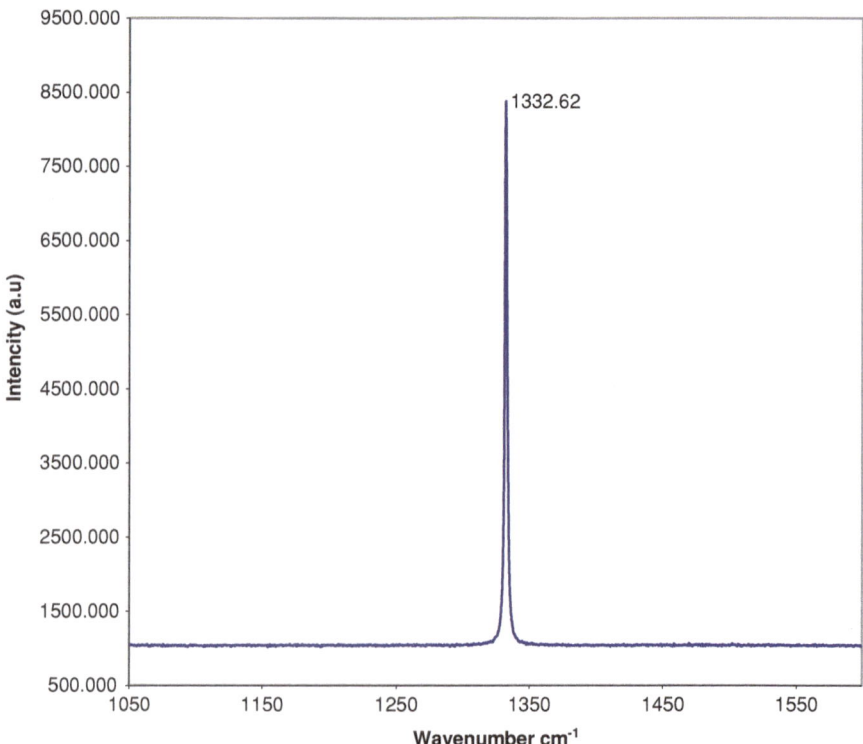

Fig. 5.9 Raman spectrum of diamond deposited using VFCVD on molybdenum wire showing the single phonon diamond peak at 1,332 cm^{-1}

Fig. 5.10 **a** Tip and cutting
edge of microdrill. **b** Cutting
edge of microdrill (*top view*)

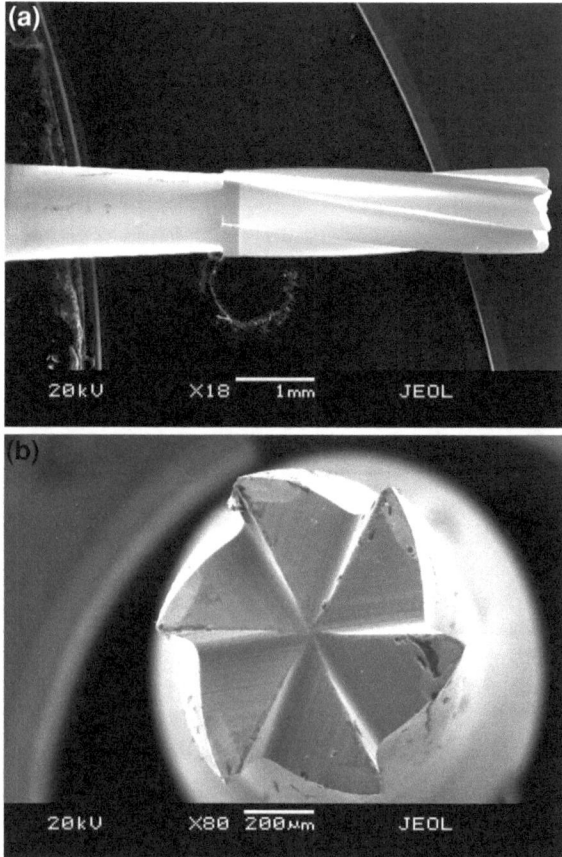

5.3 Diamond Deposition onto Microdrills

Figure 5.10a shows a SEM micrograph of a microdrill without a diamond coating.
The WC-Co cutting edges are welded onto the steel shaft (Fe–Cr). The cutting tip
is about 4 mm in length and 0.8 mm in diameter. It can be seen that the microdrill
has six cutting edges Fig. 5.10b.

Figure 5.11a, b show the SEM micrographs and the corresponding EDX spectra of the WC-Co microdrill before and after the chemical etching process. Before
etching, the EDX spectrum (Fig. 5.11a) shows peaks for cobalt, carbon and tungsten. High cobalt content inhibits diamond deposition resulting generally in the
growth of amorphous carbon phases, or at best, poor quality diamond films. The
Co diffuses to the surface regions preventing effective bonding between the substrate surface and the film coating. To improve the coating adhesion of diamond on
WC-Co tools, several approaches can be employed.

The following chemical etching procedure was used to remove the cobalt from
the bur surface. The cutting tools were ultrasonically cleaned in acetone for 10 min

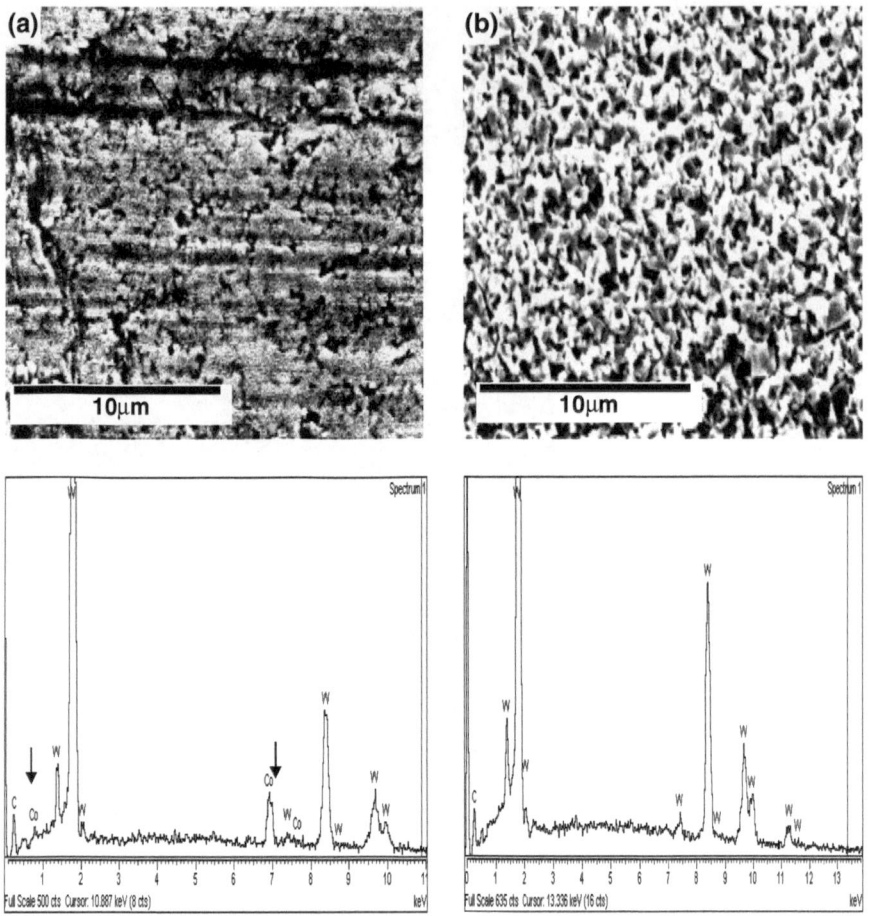

Fig. 5.11 a SEM and corresponding EDX of WC-Co microdrill before etching. **b** SEM and corresponding EDX of WC-Co microdrill after etching

in order to remove loose residues from the surface. A first etching step using Murakami's reagent ([10 g $K_3Fe(CN)_6$] + 10 g KOH + 100 ml water) was carried out for 10 min in ultrasonic bath to etch the WC substrate, followed by a rinse with distilled water. The second step etching was performed using an acid solution of hydrogen peroxide [3 ml (96 % wt.) H_2SO_4 + 88 ml (30 % w/v) (H_2O_2)] for 10 s to remove Co from the surface. The substrates were then washed again with distilled water in an ultrasonic bath.

Figure 5.11b displays the EDX spectrum after the 2-step etching process and also shows that the Co peak has disappeared. This is vital to enhancing the diamond coating adhesion. In addition, comparison of the SEM micrographs in Fig. 5.11a, b shows that the surface topography is significantly altered after etching in Murakami and H_2SO_4/H_2O_2 solutions. The etching process makes the surface much rougher but more uniform with a significant amount of etch pits. These pits can act as low energy nucleation sites for diamond crystal growth.

After substrate treatment described the microdrills were coated with CVD diamond using the modified hot filament CVD system and standard deposition conditions given in Chap. 3.

Figure 5.12 shows the SEM micrograph of a diamond coated WC-Co microdrill. It is important to note that six cutting edges of the microdrill tip were completely and uniformly coated with a polycrystalline diamond film. Thus the vertical filament arrangement and the placement of the microdrill within the coils of the filament is a good configuration for coating 3-D substrates. The diamond crystal structure and morphology were found to be uniform and adherent, as shown in Fig. 5.12a–c. Typically the crystallite sizes are of the order of 5–8 μm. The visibly adherent diamond coatings on the WC-Co microdrills consisted of mainly (111) faceted diamond crystals. The design of the filament and substrate in the reactor offer the possibility of uniformly coating even larger diameter cylindrical substrates.

Raman analysis was performed in order to evaluate the carbon-phase quality in the deposited films. The Raman spectrum in Fig. 5.13 shows a single peak at $1,335 \text{ cm}^{-1}$ at the tip of the diamond-coated microdrill, which indicates the presence of good quality diamond. The Raman spectrum also gives information about the stress in the diamond coatings. The diamond peak is shifted to a higher wave number of $1,335 \text{ cm}^{-1}$ relative to that of natural diamond peak $1,332 \text{ cm}^{-1}$ indicating that stress, which is compressive in nature, exists in the resultant coatings (Amirhaghi et al. 1999, 2001). The as-grown diamond films are often under compressive stress because diamond generally has a much lower coefficient of thermal expansion than the substrate. As the sample cools from the deposition temperature to room temperature compressive stresses develop in well-adhered films.

5.4 Performance of Diamond Coated Micro Drills

After the cutting tools were coated with diamond they were used to machine aluminium and metal alloys. The coated tools were compared with uncoated drills to distinguish them in terms of the coating adhesion and their machining behaviour. The unit has a maximum spindle speed of 500,000 revolutions/min, feed rates of between 5 and 20 μm per revolution, and cutting speeds in the range 100–200 m/min (Jackson et al. 2003). The micro machining unit is shown in Fig. 5.14 the machining centre is constructed using three principal axes each controlled using a D.C motor connected to a Motionmaster^{TM} controller. A laser light source is focused onto the rotating spindle in order to measure the speed of the cutting tool during machining. Post machining analysis was performed using a scanning electron microscope to detect wear on the flanks of the cutting edges.

After machining an aluminium alloy material, very low roughness and chipping of the diamond coated micro tool were detected. Figure 5.15 shows a typical

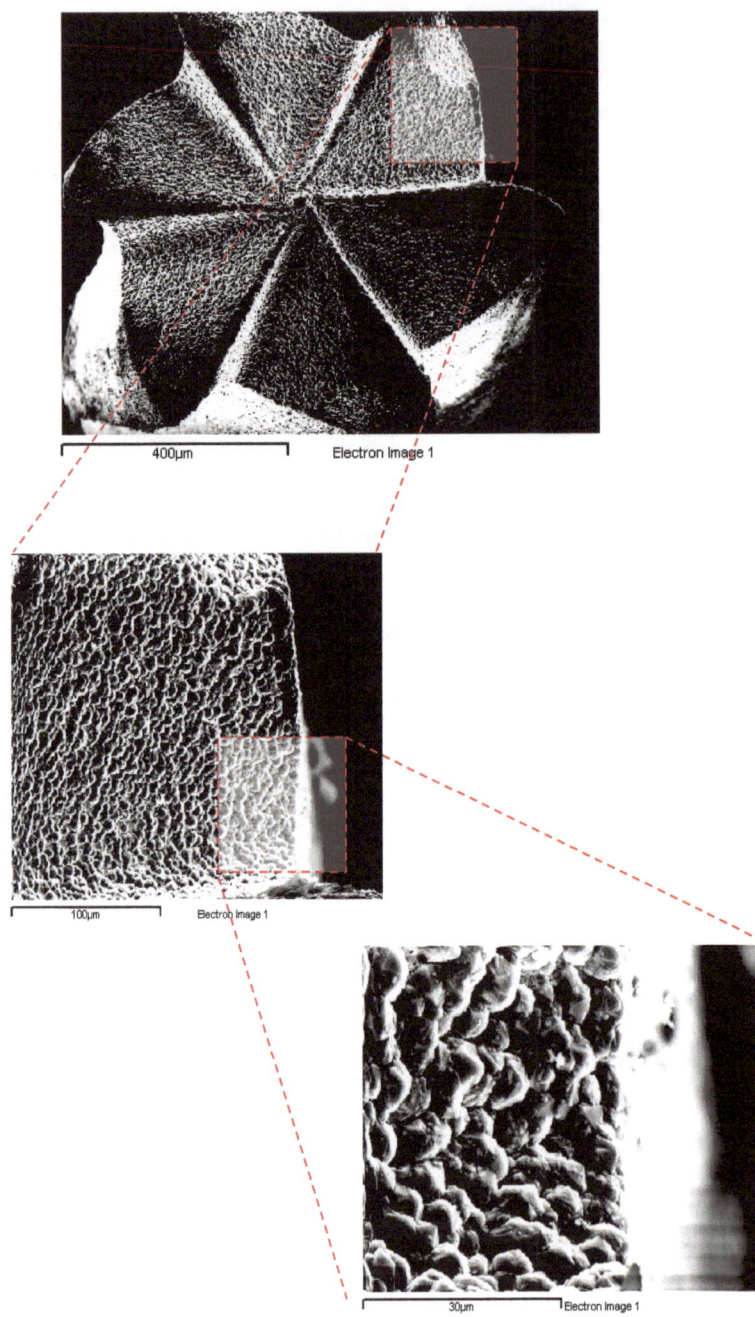

Fig. 5.12 a The cutting edge of microdrill after depositing with VFCVD diamond (*end view*), **b** the cutting edge of microdrill uniformly coated with **VF**CVD diamond, **c** the SEM of microdrill after depositing with VFCVD diamond (*close up*)

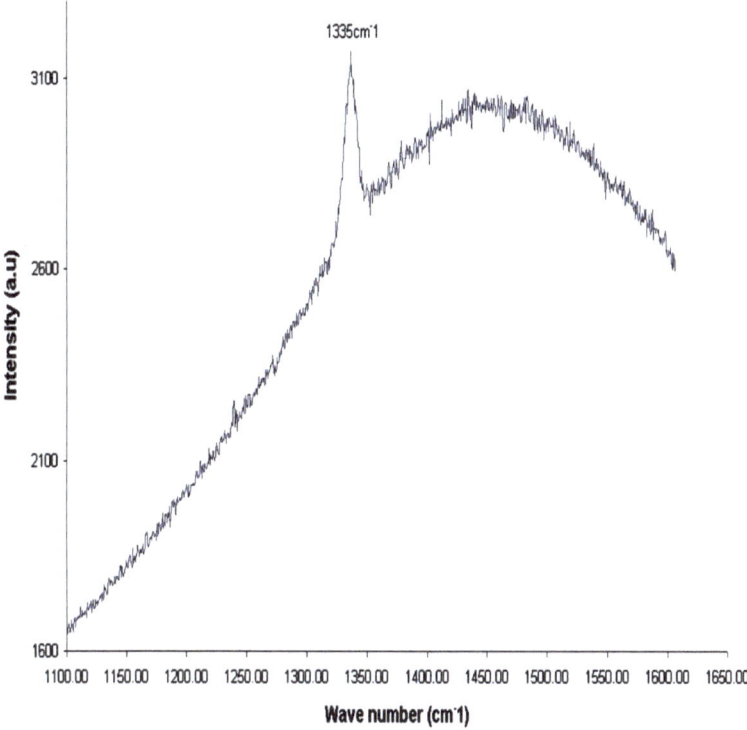

Fig. 5.13 Raman spectrum of the tip of the diamond coated WC-Co microdrill

Fig. 5.14 The micro
machining unit

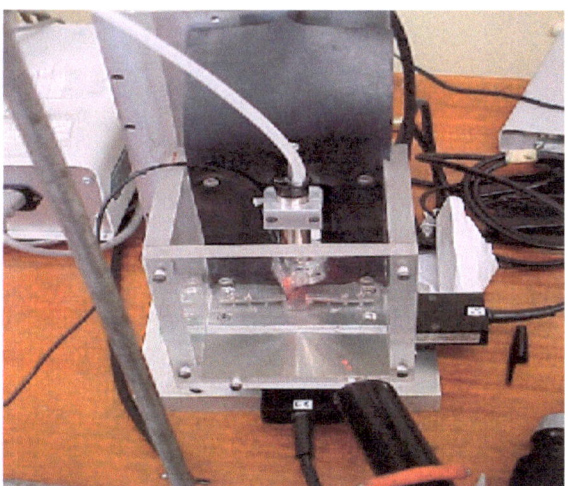

machined surface in aluminium alloy. A metal chip created from this machining operation is shown in Fig. 5.16. The chip clearly shows shear fronts separated by lamellae caused by plastic instabilities within the material generated at such high

Fig. 5.15 Aluminium alloy
material showing a machined
track produced by the
diamond coated micro cutting
tool

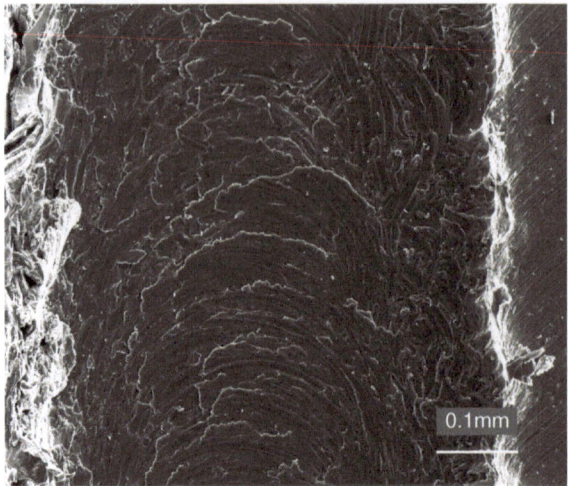

Fig. 5.16 Aluminium alloy
chip generated during a high
speed machining operation
using the diamond coated
micro cutting tool showing
attached and detached shear
fronts

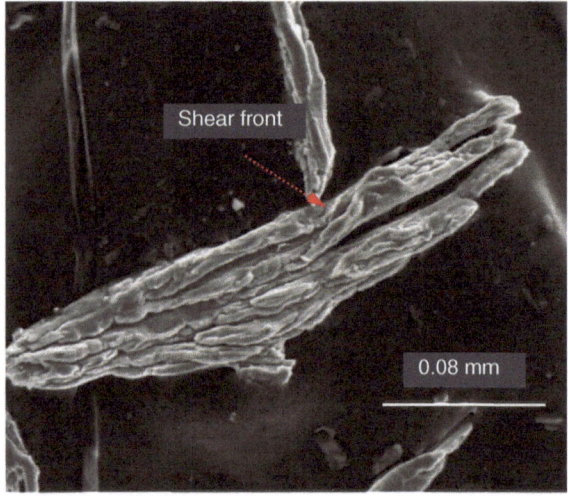

speeds. Uncoated and diamond coated tools were used to drill a series of holes in
an aluminium alloy and comparing the results. The wear of each tool was determined by examining the extent of flank wear. Uncoated tools appeared to chip at
the flank face, and diamond coated tools tended to lose individual diamonds at
the flank face. Uncoated tools drilled an average of 8,000 holes before breakdown
occurred, and the diamond-coated tools drilled an average of 24,000 holes (Sein
and Jackson 2000; Jackson et al. 2004; Polini et al. 2006; Sein et al. 2004b). It can
be concluded therefore, that diamond coated micro drills have a superior performance to uncoated microdrills.

5.5 Conclusions

The deposition of polycrystalline diamond films onto molybdenum wire and microdrills has been successfully achieved. To deposit adherent diamond films the WC-Co substrates need to be etched using a 2-step chemical etching treatment. Without etching diffusion of the Co binder occurs when the substrates are exposed to high temperature in the VFCVD reactor and hence the diamond films are less adherent to the substrate. Diamond coated microdrills have also been shown to have a superior machining performance compared to uncoated micro drills.

References

Amirhaghi S et al (1999) Growth and erosive wear performance of diamond coatings on WC substrates. Diam Relat Mater 8(2–5):845–849

Amirhaghi S et al (2001) Diamond coatings on tungsten carbide and their erosive wear properties. Surf Coat Technol 135(2–3):126–138

Ashfold MNR et al (1994) Thin film diamond by chemical vapour deposition methods. Chem Soc Rev 23(1):21–30

Jackson MJ et al (2003) Manufacture of diamond-coated cutting tools for micromachining applications. Proc Inst Mech Eng, Part L: J Mater Des Appl 217(1):77–83

Jackson MJ et al (2004) Diamond coated dental bur machining of natural and synthetic dental materials. J Mater Sci—Mater Med 15(12):1323–1331

Lu GH et al (1995) A technique for the manufacture of long hollow diamond fibres by chemical vapour deposition. J Mater Sci Lett 14(20):1448–1450

May P (1995) Synthetic diamond: Emerging CVD science and technology: Edited by Harl E. Spear and John P. Dismuhes. Pp. 663. Wiley, Chichester, 1994. £74.00 ISBN 0 4715 3589 3. Endeavour 19(1):48

May PW et al (1994) CVD diamond wires and tubes. Diam Relat Mater 3(4–6):810–813

May PW et al (1995) CVD diamond-coated fibres. Diam Relat Mater 4(5–6):794–797

May PW et al (2006) Deposition of CVD diamond onto GaN. Diam Relat Mater 15(4–8):526–530

Polini R et al (2006) Effects of Ti-and Zr-based interlayer coatings on the hot filament chemical vapour deposition of diamond on high speed steel. Thin Solid Films 494(1):116–122

Sein H et al (2004a) Enhancing nucleation density and adhesion of polycrystalline diamond films deposited by HFCVD using surface treatments on Co cemented tungsten carbide. Diam Relat Mater 13(4–8):610–615

Sein H et al (2004b) Performance and characterisation of CVD diamond coated, sintered diamond and WC–Co cutting tools for dental and micromachining applications. Thin Solid Films 447:455–461

Sein H et al (2002) Application of diamond coatings onto small dental tools. Diam Relat Mater 11(3–6):731–735

Sein H, Jackson M, Ahmed W, Rego CA (2000) New Diam Front Technol 12:1–10

Wang BB et al (2001) Theoretical analysis of ion bombardment roles in the bias-enhanced nucleation process of CVD diamond. Diam Relat Mater 10(9–10):1622–1626

Zhang GF, Buck V (2000) Influence of geometry factors of in situ dc glow discharge on the diamond nucleation in a hot-filament chemical vapor deposition system. Surf Coat Technol 132(2):256–261

Zhang GF, Buck V (2002) Lower filament temperature limit of diamond growth in a hot-filament CVD system. Surf Coat Technol 160(1):14–19

Chapter 6
Diamond Deposition on Tungsten Carbide Burs Using VFCVD

Abstract The deposition of diamond films using VFCVD onto tungsten carbide dental burs has been described. To enhance nucleation, growth and adhesion of diamond substrate was pre-treated using a Murakami etch. The structure and morphology of the diamond coated burs and uncoated burs have been compared.

Keywords Tungsten carbide (WC-Co) burs · Diamond films · Chemical vapour deposition · CVD · VFCVD · Structure and morphology

6.1 Dental Bur Substrate

VFCVD was used to deposit polycrystalline diamond films onto WC dental burs. Laboratory grade tungsten carbide (WC-Co) dental burs are shown in Fig. 6.1a (AT23 LR) as bur *A* with fine WC fine grain sizes (1 μm) 20–30 mm in length and 1.0–1.5 mm in diameter and Fig. 6.1b (HPTX 23) as bur *B* with coarse grain size (4–6 μm) 20–30 mm in length and 1.0–1.5 mm in diameter were supplied by Metro dent Ltd, UK.

6.1.1 Grains Effect on WC-Co Substrate

Two different types of dental burs Fig. 6.1a AT23 LR (*A*) and Fig. 6.1b HPTX 23 (*B*) contain 6 % Co and 94 % WC substrate with respective grain size 0.5–1 μm (*A*) and 4–6 μm (*B*) were chosen for diamond coating.

The tungsten carbide WC grain size varies even with same Co Wt % in dental burs in which WC grain varies from submicron (0.5 μm) to larger than 5 μm. Using larger grains of 4–6 μm will greatly increase the strength and toughness of the material because the larger grains interlock better. The trade-off is that larger grain materials do not offer as much resistance to wear as finer grain sized

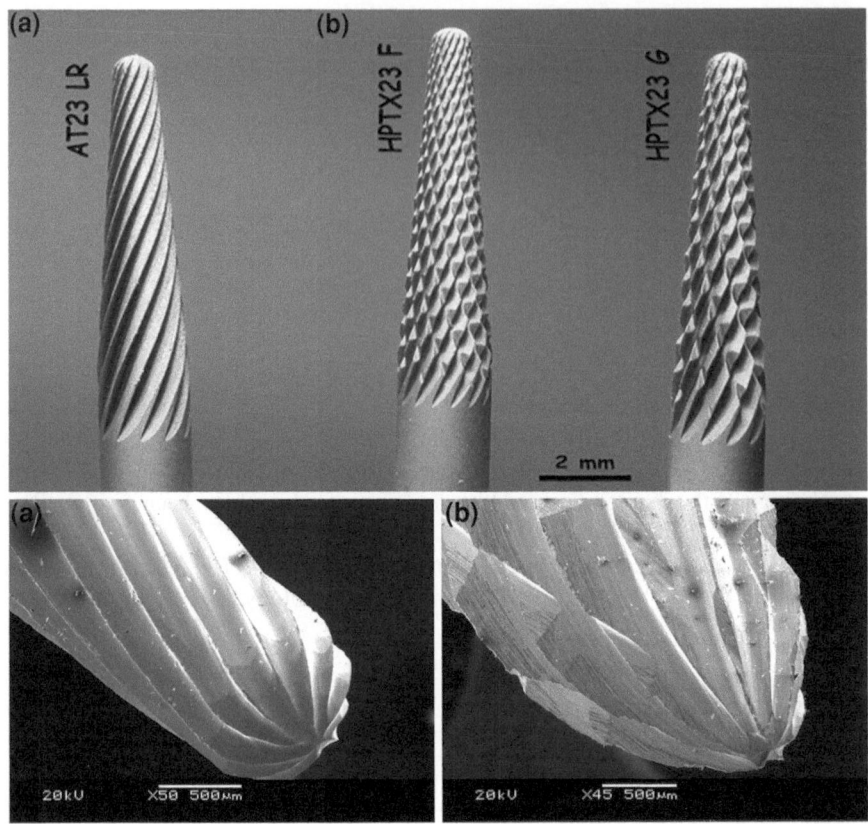

Fig. 6.1 Laboratory grade virgin burs (overview) **a** Bur *A* **AT23 LR** bur, **b** Bur *B* **HPTX 23F** bur

materials. Sub-micron materials that vary between 0.5 and 1.0 micron grain size are harder than standard grain materials with the same cobalt content. The sub-micron grains are much more uniform in size and hence give improved hardness as well as increased carbide strength (Jian et al. 2004).

6.2 Substrate Pre-treatment

Substrates, made from the first transition group such as Fe, Co, and Ni, are characterized by high dissolution and diffusion rates of carbon into those substrates—see Table 6.1 (Haynes 2012).

The graphite-diamond ratio during the VFCVD diamond deposition yields a low diamond nucleation or an amorphous carbon layer at the interface between the metal and the diamond coating. This indicates that the presence of these transition metals such as Co can be harmful to diamond deposition even at relatively low concentrations (Haubner et al. 1993).

Table 6.1 Solubility and diffusion rates of carbon atoms in metals at 900 °C

	α-Fe	γ-Fe	Co	Ni
Solubility of carbon (wt %)	1.3	1.3	0.1	0.2
Carbon diffusion rate (cm²/s)	2.35×10^{-6}	1.75×10^{-8}	2.46×10^{-8}	1.4×10^{-8}

Owing to the absence of a stable carbide layer, the incubation time required to form diamond is higher and depends on substrate thickness. In order to improve the adhesion between diamond and WC substrates it is necessary to etch away the surface Co and prepare the surface for subsequent diamond growth. In particular the presence of cobalt (Co) binder, which provides additional toughness to the tool but adversely affects diamond adhesion. The adhesion strength to diamond films is relatively poor, and can lead to catastrophic failure of coating in metal cutting (Babaev and Guseva 1999; Guseva et al. 1997; Sein et al. 2003). The Co binder can also suppress diamond growth favouring the formation of non-diamond carbon phases resulting in poor adhesion between the diamond coating and the substrate (Endler et al. 1999; Sein et al. 2003).

Most importantly, it is difficult to deposit adherent diamond onto untreated WC-Co substrates. Poor adhesion can be related to the cobalt binder that is present to increase the toughness of the tool however it suppresses diamond nucleation and causes deterioration of diamond film adhesion. To eliminate this problem, it is usual to pre-treat the WC-Co surface prior to VFCVD diamond deposition. Various approaches have been used to suppress the influence of Co and to improve adhesion. Therefore, a substrate pre-treatment, for reducing the surface Co concentration and achieving a proper interface roughness, will enhance the surface readily available for coating process (Polini et al. 2002). Emergence of cobalt content on the surface of substrate inhibits diamond deposition resulting generally in the growth of amorphous carbon phases also preventing to grow adhered diamond film and diffuse Co can disturb the effective bonding between the substrate surface and the film coating (Amirhaghi et al. 1999, 2001; Polini et al. 2003; Polini 2006). The cutting tools were ultrasonically cleaned in acetone for 10 min in order to remove loose residues from the surface. A first etching step using Murakami's reagent ([10 g $K_3Fe(CN)_6$] + 10 g KOH + 100 ml water) was carried out for 10 min in ultrasonic bath to etch the WC substrate, followed by a rinse with distilled water. The second step etching was performed using an acid solution of hydrogen peroxide [3 ml (96 % wt.), H_2SO_4 + 88 ml (30 % w/v) (H_2O_2)] for 10 s to remove Co from the surface. The substrates were then washed again with distilled water in an ultrasonic bath (Sein et al. 2002; Kulisch and Popov 2006; Popov et al. 2004, 2006). The surface roughness depends on the original WC-Co grains sizes. Two types of WC-Co are shown in Fig. 6.2a fine grain and b coarse grain, after 2 step etching process. It is expected that the final film geometry of the VFCVD diamond deposited will be highly dependent on the surface characteristics of the etched substrates. The resulting VFCVD diamond can be seen resulting in respective coarse and fine grains.

As mentioned earlier, a material with a large WC particle size and a high percentage the material has a high shock resistance resulting in high impact strength. When the material has fine grains of WC and a lower Co content then it is harder and has better wear resistance.

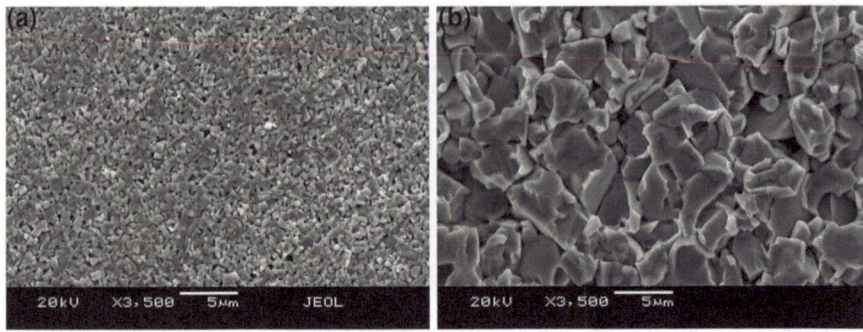

Fig. 6.2 **a** Fine grained, Bur A and **b** coarse grained, Bur B

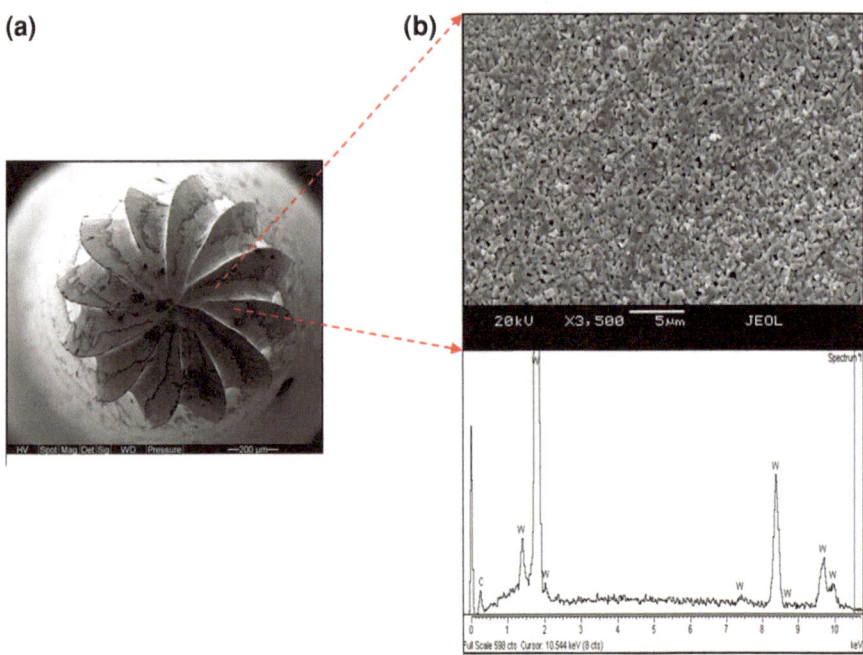

Fig. 6.3 **a** Laboratory used WC-Co dental bur *A* AT23LR **b** Dental bur surface after etching (chemical treatment)

Figures 6.3 and 6.4 display the EDX spectrum after the 2-step etching process. It shows that the Co peak has disappeared. This is a vital factor the diamond coating adhesion. In addition, comparison of the SEM micrographs in Fig. 6.3a, b for bur *A* and Fig. 6.4a, b for bur *B* shows that the surface topography is significantly changed and surface Co also disappeared after etching in Murakami and H_2SO_4/H_2O_2 solutions. The etching process makes the surface much rougher with a significant amount of etch pits which can be useful for low energy nucleation sites and enhanced diamond crystal growth.

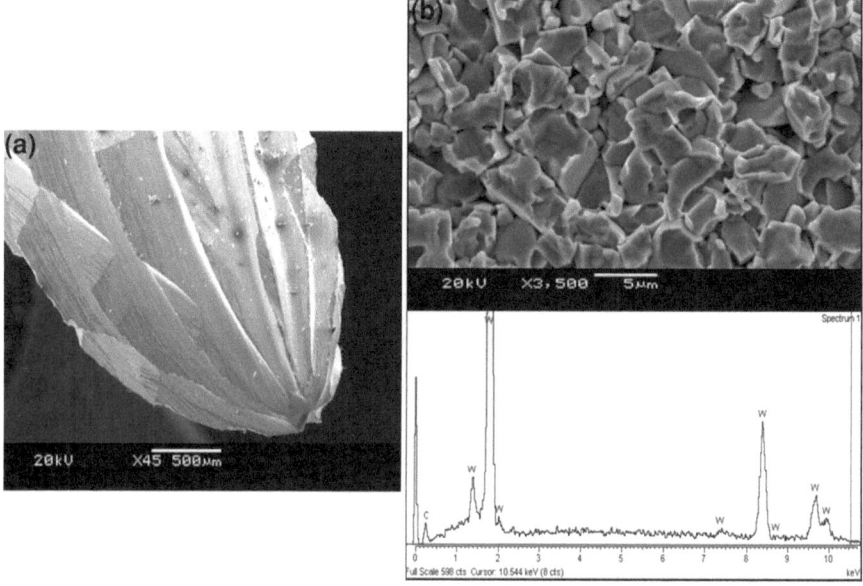

Fig. 6.4 **a** Laboratory used WC-Co dental bur *B* HPTX23F **b** Dental bur surface after etching (chemical treatment)

6.3   Diamond Deposition Dental Bur Without Substrate Pre-treatment

VFCVD diamond was also coated on an un-etched WC-Co substrate. Figure 6.5a shows SEM and EDX of diamond deposited onto WC-Co substrates without etching treatment for 5 h. Polycrystalline diamond films have been deposited. However, it is evident that cobalt has diffused to the surface of the diamond films. Figure 6.5a shows those tiny sub-micron size Co crystals on diamond films that deposited on untreated substrate (without removal of surface Co). EDX spectra show that trace amount of Co elements on the surface. Poor adhesion of diamond film is evident from the delaminated film on the surface as shown in Fig. 6.5b. The diffusion of cobalt is enhanced at the substrate temperatures employed at the VFCVD conditions applied.

6.4   VFCVD Diamond Deposition

Tantalum wire of 0.5 mm in diameter and 12–14 cm in length was used as the hot filament. The filament was mounted vertically with the dental bur held in between the filament coils, as opposed to the horizontal position used in the conventional HFCVD system. To ensure uniform coating the dental bur was positioned centrally

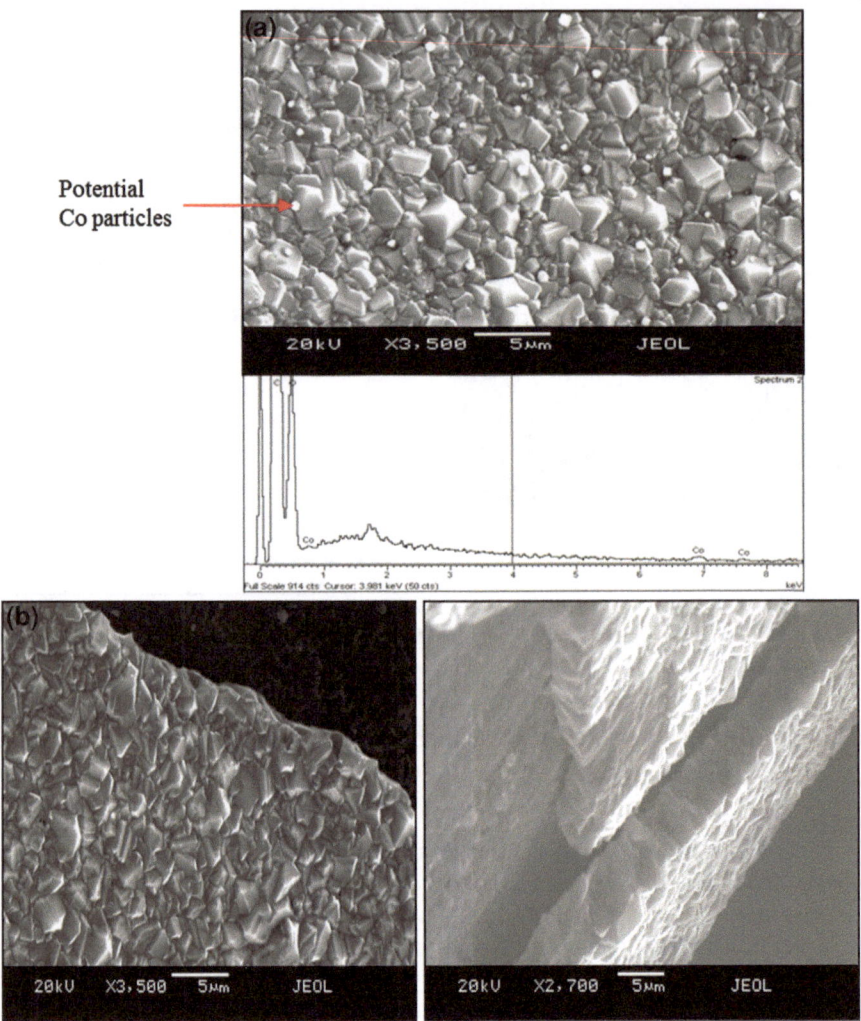

Potential
Co particles

Fig. 6.5 a Co appeared on surface of diamond films with EDX **b** Delaminated diamond film which peeled off from substrate surface

and coaxially within the coils of the filament. A schematic diagram of the VFCVD system is presented in Fig. 6.6 has designed for dental bur or micro drill with similar diameter. The new vertical filament arrangement used in the VFCVD system should enhance the thermal distribution, ensure uniform coating, increase growth rates and produce higher nucleation densities.

Diamond coatings were deposited using experimental conditions given in Table 3.2 and described in Chap. 3. After 12 h deposition VFCVD diamond are uniformly coated on dental bur. Figure 6.7 shows the SEM micrograph of a VFCVD diamond coated dental bur **A** (AT 23LR) at the cutting edge. The film is

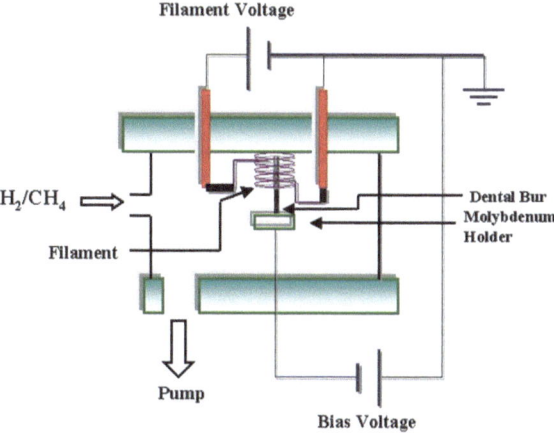

Fig. 6.6 Schematic of typical VFCVD system

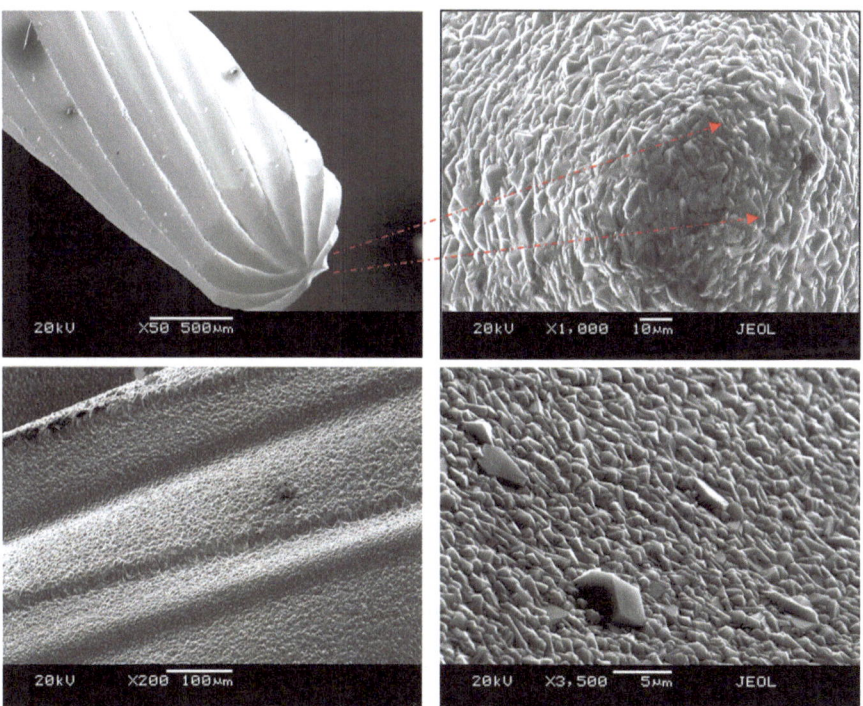

Fig. 6.7 (111) faceted octahedral VFCVD diamond film on dental bur **A** AT 23 LR

homogeneous with uniform diamond crystal sizes. Typically the crystal sizes are in the range of 3–5 μm on the middle of bur and larger crystal size 6–8 μm of diamond grows at bur tips, which have more energy in the centre of filament due

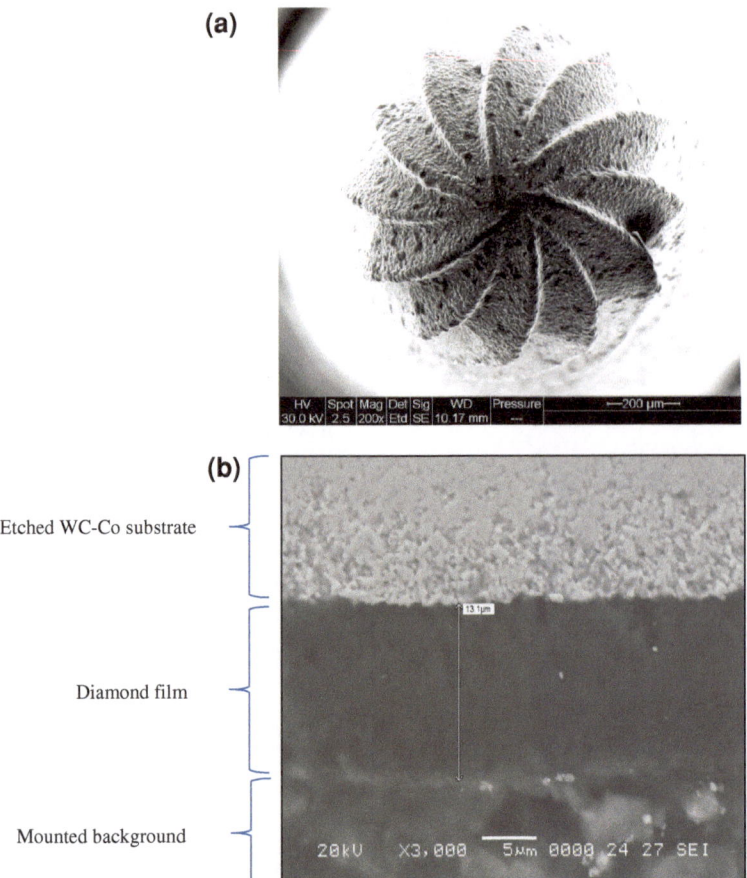

(a)

(b)

Etched WC-Co substrate

Diamond film

Mounted background

Fig. 6.8 **a** topical view of diamond coated dental bur *A* **b** Thickness of diamond on dental bur *A*

to higher temperatures. As expected the surface morphology is rough making the dental burs extremely desirable for abrasive applications.

Surface morphology of predominately (111) faceted octahedral shape diamond films was obtained after deposition for 12 h for bur *A*. Figure 6.8a shows a topical view of dental bur with uniform diamond growth on the cutting edge and cross section of 13 µm thick diamond is shown in Fig. 6.8b. Figure 6.8a topical view of dental bur *A*.

Raman analysis was performed in order to evaluate the quality and stress imparted in CVD diamond films. The Raman spectrum shows that at the tip, centre and end of cutting tool a single sharp peak at 1,336, 1,337 and 1,337 cm^{-1} respectively was observed for different positions (Fig. 6.9). The result of Raman analysis on WC-Co substrates at several different locations on the tool shows indications of compressive stress in the coating (Tsang et al. 1997).

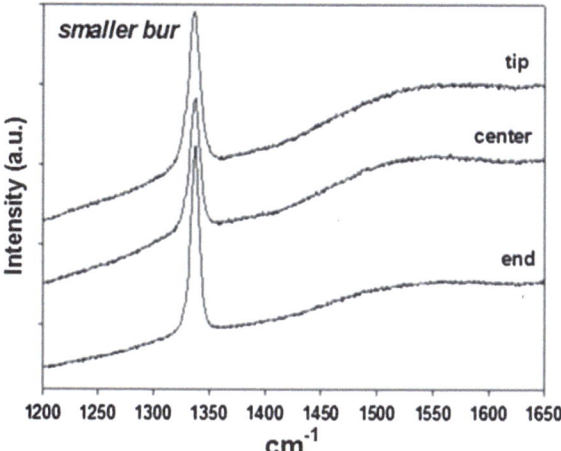

Fig. 6.9 Raman spectra of Bur *A* AT 23LR

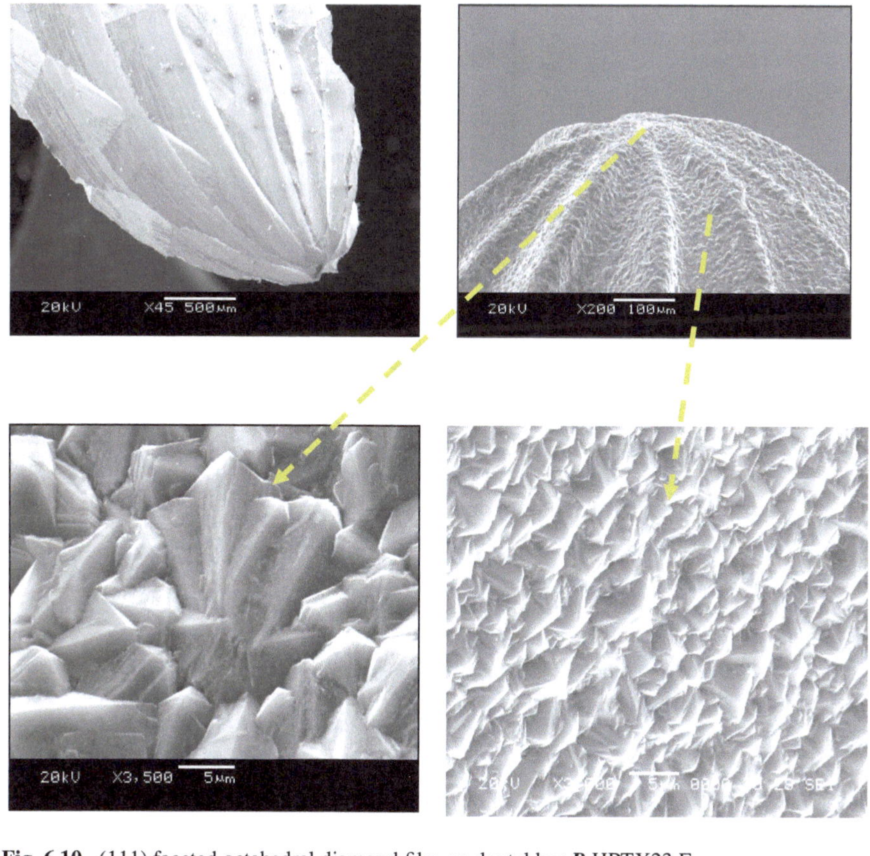

Fig. 6.10 (111) faceted octahedral diamond film on dental bur *B* HPTX23 F

Fig. 6.11 Thickness of
diamond on dental bur **B**

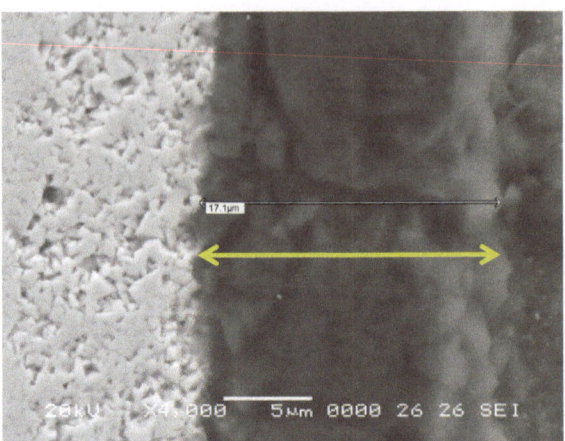

Fig. 6.12 Raman spectra
of dental bur **B** HPTX23
F. Diamond peak position:
tip: $1,337 \pm 1$ cm^{-1}; center:
$1,337 \pm 1$ cm^{-1} end:
$1,336 \pm 1.5$ cm^{-1}

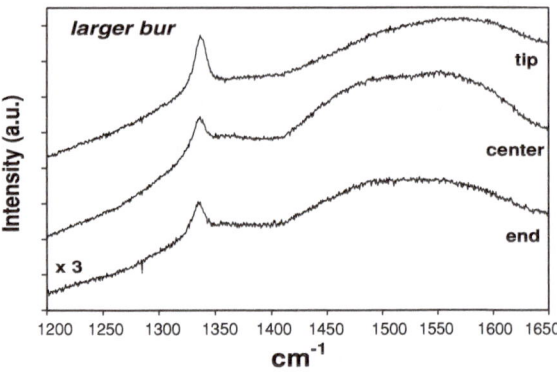

Figure 6.10 shows the SEM micrograph of a VFCVD diamond coated dental
bur **B** (HPTX23 F) at the cutting edge. The film is homogeneous with uniform diamond crystal sizes. Diamond layers were deposited on pre-treated WC-Co dental
burs. WC has a much higher affinity for diamond nucleation than molybdenum
due to readily available WC layer on the surface of dental bur since Murakami
solution and acid etching had etched cobalt away from the surface. The diamond
deposition rate on WC-Co dental bur coated under identical condition is about
1 μm/h, thus after 15 h deposition time, the film thickness was measured to be
15–17-μm (Fig. 6.11).

The schematic of typical cross section of dental bur has shown in Fig. 6.11
shown the thickness of dental bur is 17.1 μm of diamond film on etched WC-Co
dental bur surface.

The Raman spectrum in Fig. 6.12 also gives an indication about the stress in the
diamond coating. The diamond peak is shifted to a higher wave number of magnitude such as $1,336, 1,337$ cm^{-1} than that normally experienced in an unstressed
coating where the natural diamond peak occurs at $1,332$ cm^{-1} therefore it is

indicating that the stress is compressive. The Raman spectra shown in Fig. 6.12 also show that at the tip, centre and end of dental bur **B** HPTX23F produced a single sharp peak at 1,337, 1,337 and 1,336 cm^{-1} respectively indicating compressive stress in the diamond films.

References

Amirhaghi S et al (1999) Growth and erosive wear performance of diamond coatings on WC substrates. Diam Relat Mater 8(2–5):845–849

Amirhaghi S et al (2001) Diamond coatings on tungsten carbide and their erosive wear properties. Surf Coat Technol 135(2–3):126–138

Babaev VG, Guseva MB (1999) Ion-assisted condensation of carbon. Carbyne and carbynoid structures. Springer, Berlin, pp 159–171

Endler I et al (1999) Preparation and wear behaviour of woodworking tools coated with superhard layers. Diam Relat Mater 8(2):834–839

Guseva MB et al (1997) Phase transition in C: N films under shock wave compression. Diamond based composites. Springer, Berlin, pp 161–169

Haubner R et al (1993) Diamond deposition on chromium, cobalt and nickel substrates by microwave plasma chemical vapour deposition. Diam Relat Mater 2(12):1505–1515

Haynes WM (2012) CRC handbook of chemistry and physics. CRC Press, Boca Raton

Jian XG et al (2004) Study on the effects of substrate grain size on diamond thin films deposited on tungsten carbide substrates. Key Eng Mater 274:1137–1142

Kulisch W, Popov C (2006) On the growth mechanisms of nanocrystalline diamond films. Phys Status Solidi A 203(2):203–219

Polini R (2006) Chemically vapour deposited diamond coatings on cemented tungsten carbides: substrate pretreatments, adhesion and cutting performance. Thin Solid Films 515(1):4–13

Polini R et al (2002) Effect of substrate grain size and surface treatments on the cutting properties of diamond coated Co-cemented tungsten carbide tools. Diam Relat Mater 11(3):726–730

Polini R et al (2003) Dry turning of alumina/aluminum composites with CVD diamond coated Co-cemented tungsten carbide tools. Surf Coat Technol 166(2):127–134

Popov C et al (2004) Growth and characterization of nanocrystalline diamond/amorphous carbon composite films prepared by MWCVD. Diam Relat Mater 13(4):1371–1376

Popov C et al (2006) Nanocrystalline diamond/amorphous carbon composite films for applications in tribology, optics and biomedicine. Thin Solid Films 494(1):92–97

Sein H et al (2002) Application of diamond coatings onto small dental tools. Diam Relat Mater 11(3–6):731–735

Sein H et al (2003) Stress distribution in diamond films grown on cemented WC-Co dental burs using modified hot-filament CVD. Surf Coat Technol 163:196–202

Tsang RS et al (1997) Examination of the effects of nitrogen on the CVD diamond growth mechanism using in situ molecular beam mass spectrometry. Diam Relat Mater 6(2):247–254

Chapter 7
Controlling Structure and Morphology

Abstract The performance of the burs and tools coated with diamond depend on the properties of the films deposited including the structure, morphology and adhesion onto the substrate material. These film characteristics are in turn controlled by a number of factors including the VFCVD reactor parameters such as substrate temperature, substrate bias, pressure, methane concentration, plasma characteristics and affinity of the substrate for diamond nucleation. In this chapter the effects of process parameters on the structure and morphology of the films deposited and therefore control of the film properties.

Keywords VFCVD · Structure · Morphology · Diamond films · Nucleations

7.1 Introduction

Numerous studies have been carried out onto flat silicon substrates and the general trends are well established (Sein et al. 2004a, b; Polini et al. 2004; Cock et al. 1994; Matsumoto et al. 1982; May et al. 2010). However, deposition of adherent high quality diamond onto substrates such as cemented carbides, stainless steel and various metal alloys containing transition elements present a considerable challenge due to poor adhesion and low nucleation density (Mankelevich and May 2008; Murakawa and Takeuchi 1991; Deuerler et al. 1996; Xianglin 2008).

As mentioned earlier feed gases methane and hydrogen are commonly used to grow diamond films of high quality. These films have been found to be rough and brittle. For smooth, hard and tough films a different solution is needed for a number of applications especially involving metal substrates where the films tend to delaminate due to large thermally induced residual stresses. For many applications the films need to adhere strongly to the surface and have the ability to plastically deform without fracturing. Hence, nanostructured diamond films with small diamond grains from about 5 to 100 nm imbedded in an amorphous carbon phase could be of significant value. Even though these films have a hardness of about 80 % of natural diamond, are smoother and have higher fracture toughness. Several different options are available using a combination of gases, process parameters and combined methods.

Nucleation of diamond is an important step in the growth of diamond thin films, because it strongly influences the diamond growth process, film quality and morphology (Butler et al. 2009; Filik et al. 2006; Schneider et al. 2010; Amirhaghi et al. 2001). Generally, seeding or abrading with diamond powder or immersing in diamond paste containing small crystallites processed in an ultrasonic bath enhances nucleation. The most promising in situ method for diamond nucleation enhancement is negative substrate biasing during the initial stage of deposition (Wang et al. 1999, 2001). Some work has been done on negative BEN for flat WC-Co inserts using microwave plasma CVD. In this chapter the results obtained on negative BEN of diamond films deposited onto complex 3-D shaped rotary cutting tools are described. A modified vertical hot filament was employed and the key process parameters, which affect the structure and morphology have also been investigated.

7.2 Effects of Temperature

The substrate temperature has a profound effect the crystal size and morphology of the diamond films. Using the vertical arrangement the dental burs were placed concentrically within the filament coils. With this configuration the temperature varies significantly from the middle to the ends of the filament. This results in variable temperature at the substrate. The structure and the morphology of diamond films are thus affected significantly.

Therefore, an analysis of temperature distribution along the coiled filament is required. Filament temperature measurements were carried out using a two-colour optical pyrometer at various positions along the filament. The position of the bur within the filament is defined in Fig. 7.1.

There is a significant temperature gradient along the filament and that temperature is highest at the centre of the filament (Fig. 7.2).

The bur substrate temperature was also measured parallel to the positions A, B, and C using a K-type thermocouple. These measurements show that the substrate temperature varies according to the position within the filament from 950 °C at position A to 840 °C at position C.

Figure 7.3 shows SEM micrographs of the as-grown diamond film at the tip, middle and base of the WC-Co dental bur (cf. positions A, B and C, respectively in Fig. 7.1). It can be seen that regardless of the position within the filament the films are polycrystalline with predominantly (111) faceted octahedral diamond. It is evident that the films are continuous, uniform and with good coverage. In addition, the grooves and the cutting edges were successfully covered with the diamond coating. It is significant that the diamond morphology does not change significantly between positions A, B and C on the coated bur. However, the

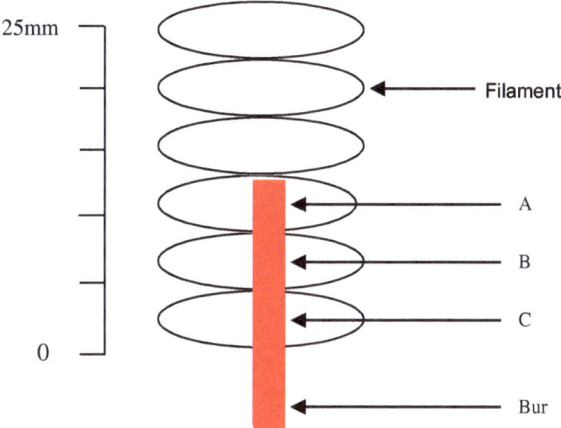

Fig. 7.1 Schematic diagram of the bur situated within the filament defining positions *A*, *B* and *C* on the bur

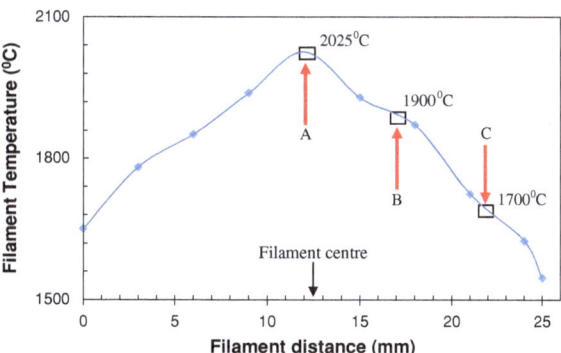

Fig. 7.2 Temperature of the filament as a function of the dimensions

crystal size does vary with position. The crystal sizes are 8–10, 3–5 and 2–4 μm, for positions A, B and C, respectively. The variation in the average crystallite size along the length of the bur is most likely caused by the variation in substrate temperature, with larger crystal sizes growing at higher substrate temperatures.

Figure 7.4 show the Raman spectra of the as-grown diamond film coating at positions A, B and C. It is evident from the spectra that all the films display a rather broad band centred at ~1,500 cm^{-1}. These broad bands indicate the presence of an amorphous carbon phase in the coating. The intensity of the Raman diamond peak relative to the broad bands is almost identical in all three positions, which indicates that the quality of the diamond is similar at the three positions on the bur. However, the peak energies are slightly different from one another. The peaks at positions A, B and C on the bur are at 1,338, 1,336, and 1,335 cm^{-1}, respectively.

The variation in the peak position can be used to calculate stress in the films. Ager and Drory (1993) investigated residual biaxial stress in diamond films grown on a titanium alloy by Raman spectroscopy and developed a general model, which

Fig. 7.3 The effect of bur position in the filament coil on the diamond crystallite size (position within the filament) is indicated on the micrograph

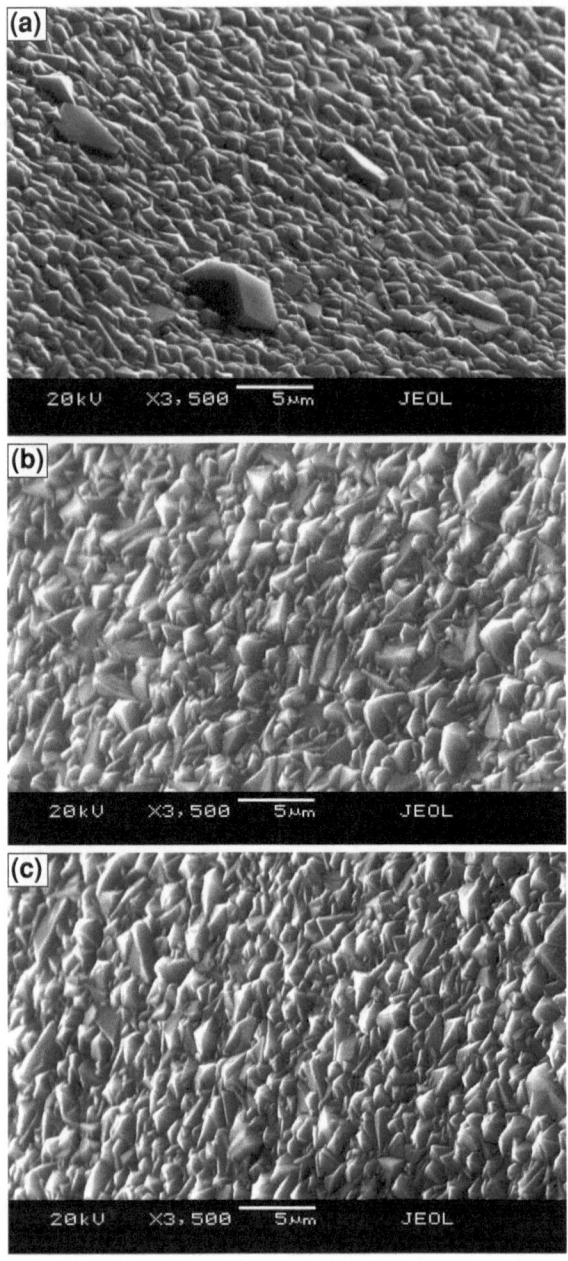

describes quantitatively the relations between singlet and doublet phonon scattering and the biaxial stress σ as follows (Sein et al. 2004a, b; Polini et al. 2010; Cabral et al. 2006)

$$\sigma = -0.567(v_m - v_0) \text{ (GPa)} \tag{6.1}$$

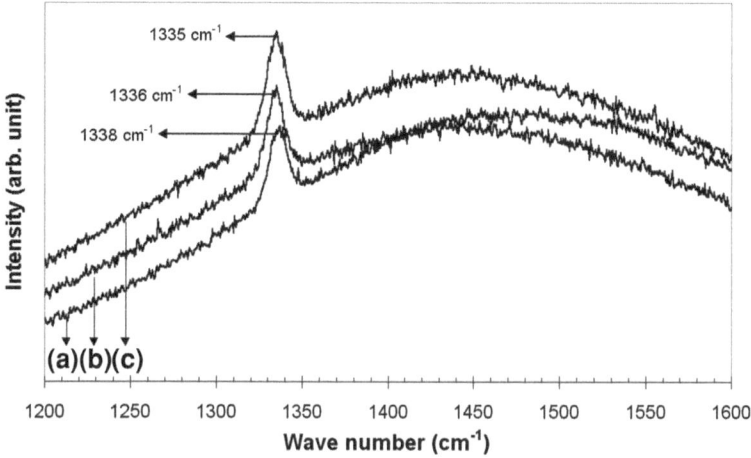

Fig. 7.4   Raman spectra at positions *A*, *B* and *C* corresponding to SEM's in Fig. 7.3

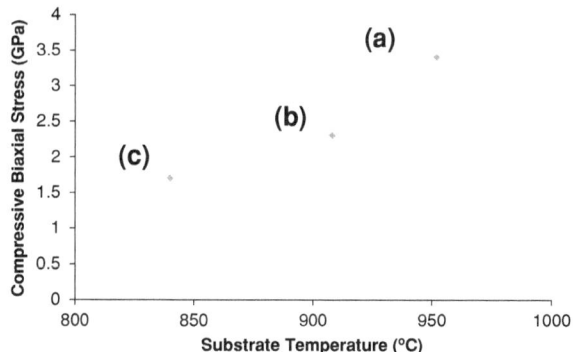

Fig. 7.5   Effect of substrate temperature on the biaxial stress at positions **a**, **b** and **c** on the dental bur

where, v_o is 1,332 cm^{-1} and v_s is the observed maximum of the singlet phonon in the spectrum. Film stress at positions A, B and C were calculated to be -3.4, -2.3 and -1.7 GPa, in compression, respectively. Figure 7.5 shows the residual stress in the diamond film grown on the cemented WC-Co dental bur as a function of the substrate temperature. Highest stress seems to occur where the deposition temperature is greatest.

Work by Wang et al. has shown Raman shift of diamond coating can be related to the cobalt content in the WC substrate (Wang et al. 1999, 2001). Higher cobalt (10 %) concentration of WC-Co dental bur shows compressive stress in the diamond coating. However, in this work it is clear that variations in the film thickness and crystal sizes evident from Fig. 7.3 are due to variation in the deposition temperature at various positions on the bur (Fig. 7.2).

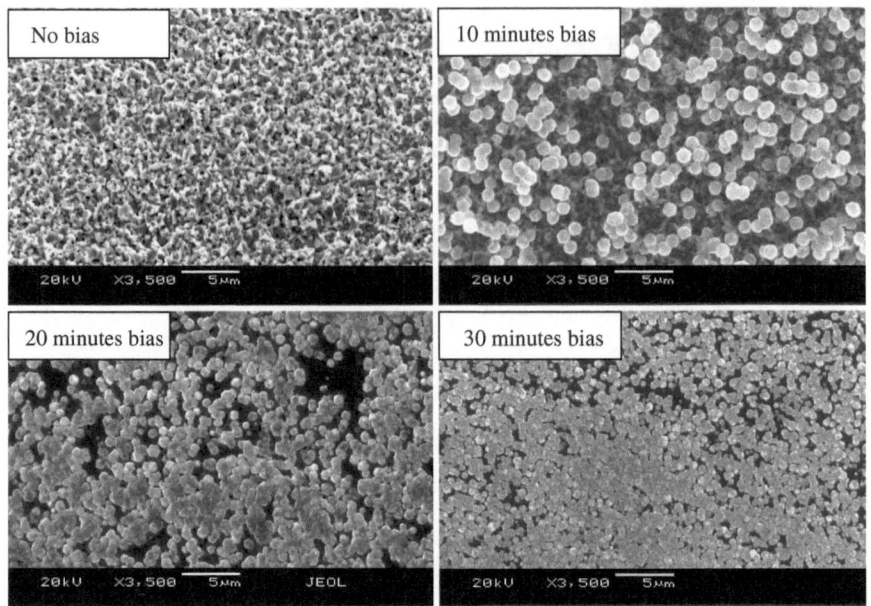

Fig. 7.6 SEMs showing the effect of bias time on nucleation density on the WC bur

7.3 Effects of Negative Bias Enhanced Nucleation on the Dental Bur

A negative bias voltage up to -300 V was applied to the substrate relative to the filament using 3 %CH$_4$ in H$_2$ at 20 Torr. This produced emission currents up to 200 mA. The nucleation times used were between 10–30 min.

The nucleation density of diamond was determined from SEMs. Figures 7.6 and 7.7 shows the effects of bias time on the nucleation density. It is evident that as the bias time increased the nucleation density also increases. The highest nucleation density was calculated to be 9×10^9 cm^{-2} for a bias time of 30 min. The calculation was done visually by counting the number of nuclei in a 1 cm^2 area on a SEM micrograph at several places and multiplied by an appropriate magnification factor to obtain the nucleation densities; this is then averaged to obtain the nucleation density. At a bias time of 10 min the nucleation density obtained was 5×10^8 cm^{-2}.

Figure 7.8 shows the variation of the emission current as a function of the negative bias applied to the substrate holder where the bias emission current increased rapidly after -180 V (Schneider et al. 2010). Wang et al. also reported that an increase in the emission current produced higher nucleation densities Sein et al. (2004a, b). Since, the bias voltage and emission current are related, the enhancement of the nucleation density cannot be attributed to solely ion bombardment or electron emission of the diamond coated molybdenum substrate holder, but may be a

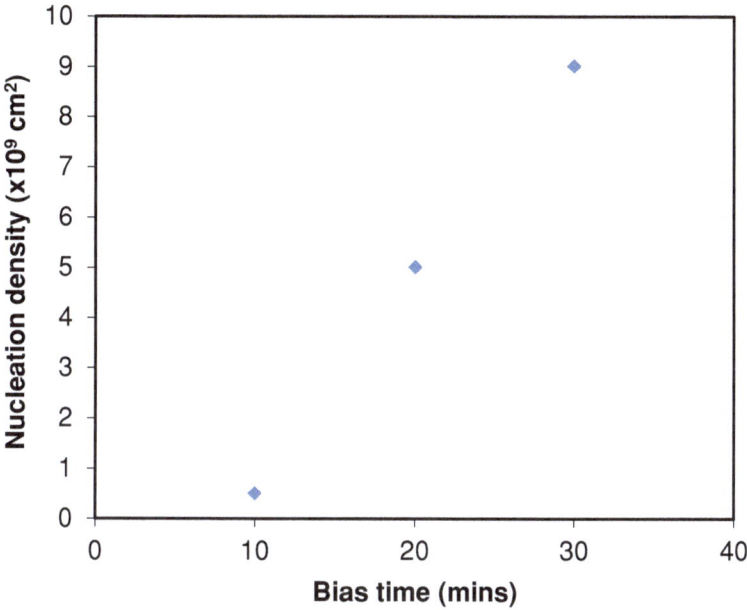

Fig. 7.7 Effect of biasing time on the nucleation density

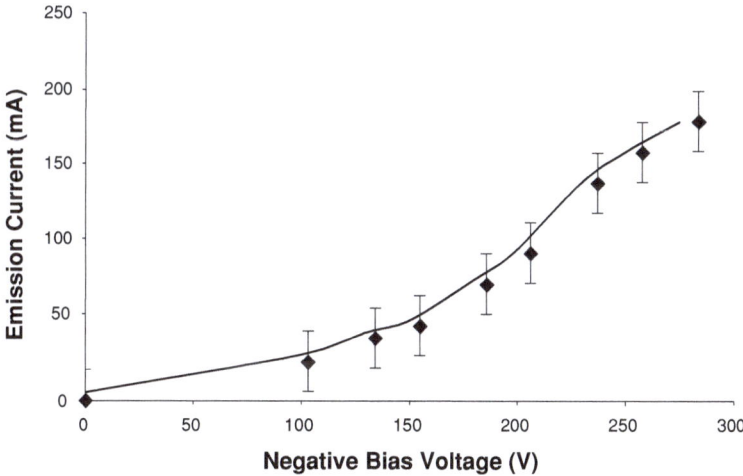

Fig. 7.8 Effect of negative bias voltage on the emission current

combination of these mechanisms Sein et al. (2004a, b). Our result was purely based on negatively bias enhanced nucleation related to the grounded filament. However, Polo et al. reported that very low electric biasing current values (μA) were detected for applied substrate biases voltages either positive or negative. Furthermore, increasing negative biases of up to -200 V resulted in value of nucleation density

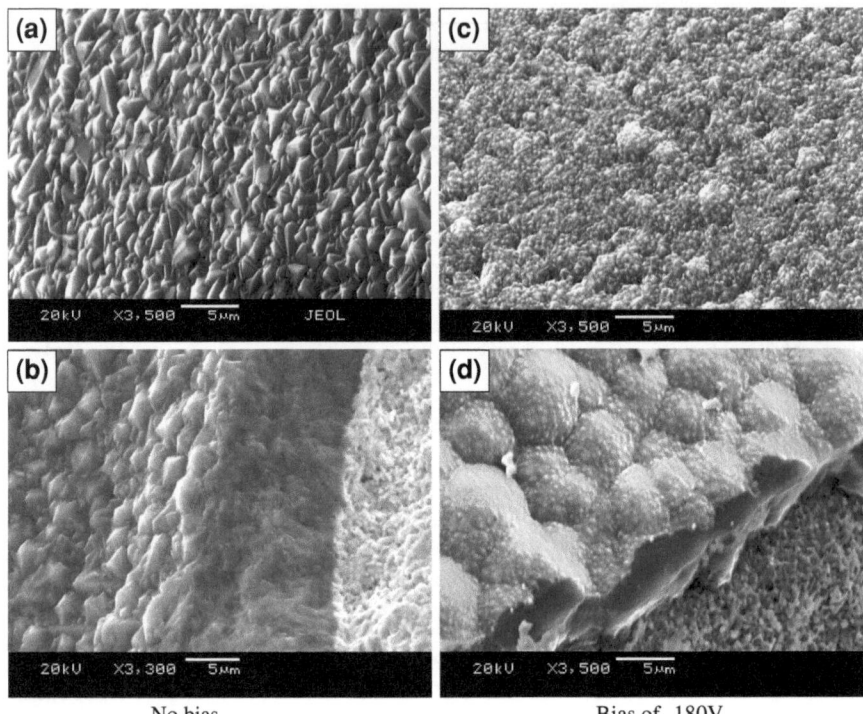

No bias   Bias of -180V

Fig. 7.9 SEMs showing **a** the surface and **b** the cross section of the diamond film without bias and **c** the surface and **d** cross section of the diamond film with -180 V negative substrate bias

is similar to that obtained with positively bias enhanced nucleation. In contrast, an application of negative bias applied to the substrate at -250 V resulted in (10^{10} cm^{-2}) maximum values of nucleation density. The enhancement in the nucleation density could be attributed to the electron current from the filament by increasing the decomposition of H_2 and CH_4 (Sein et al. 2004a, b). The increase in the nucleation density is as expected since negatively biasing the substrate increases the rate of ion bombardment into the surface creating greater numbers and density of nucleation sites which effects the surface structure. (Fig. 7.9).

Therefore, the greater the density of nucleation sites the higher the expected nucleation density. Kamiya et al. reported that reproducibility of the experiment was poor and that no definite trend in the nucleation density could be found with respect to different bias conditions (Wang et al. 2001).

Figure 7.10 shows SEMs of diamond on the cutting tool edges with negative biasing (a) and (b) and without any bias treatment (c) and (d). It is evident that the cutting edges are uniformly coated in both cases. However, when biasing is employed there is a considerable reduction of average crystal size, therefore negative BEN creates more nucleation sites for diamond growth resulting in smaller average crystallite sizes.

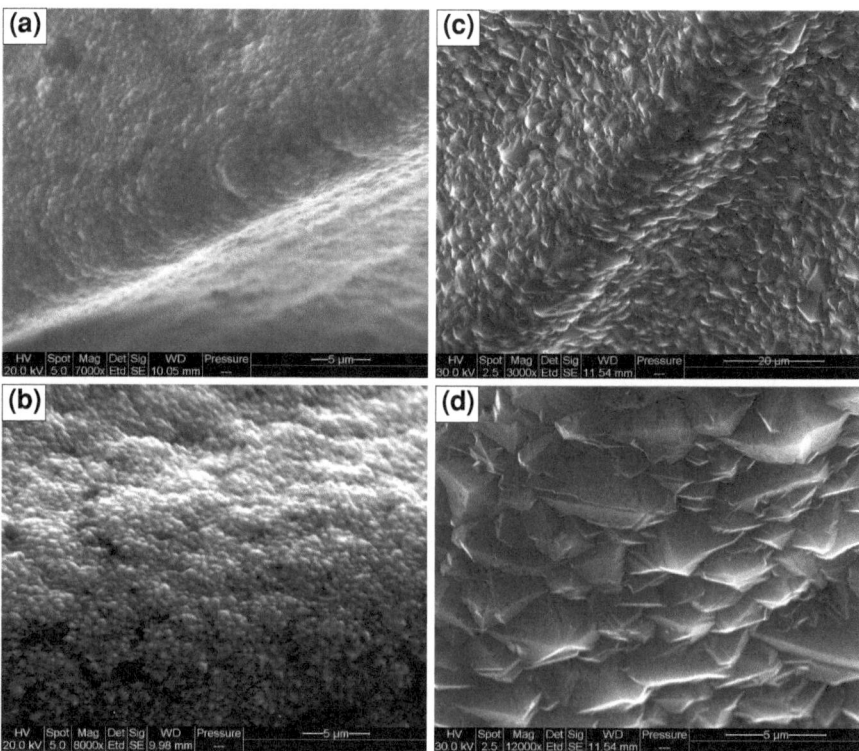

Fig. 7.10 SEMs showing cutting edges bias of the bur **a** and **b** with −180 V negative substrate and **c** and **d** without

The negatively bias assisted and diamond coated WC-Co dental burs were tested with human teeth in order to observe the adhesive strength of diamond particles on the surface. Results of these experiments are presented in Chap. 8.

7.4 Influence of Process Parameters on VFCVD Diamond Deposition

The effects of surface pre-treatment on deposition of diamond films have been investigated. Other essential factors, which affects the diamond deposition process is the deposition process parameters. The effects of process parameters such as deposition pressure, gas concentration (especially precursor methane gas), Filament voltage (DC power to filament), filament position and distance from substrate, substrate bias time and substrate temperature respect to filament temperature have been investigated and results were presented in following sections. In order to study the effects of process parameters on the diamond growth only one parameter was changed, whilst other parameters were fixed at their standard value.

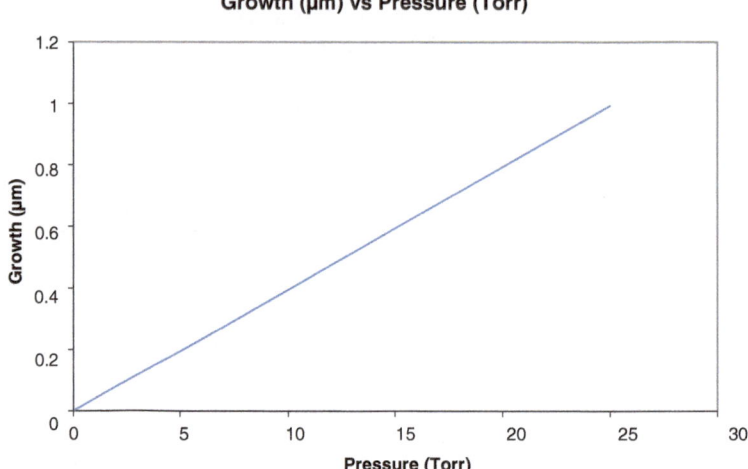

Fig. 7.11   Growth (μm) as a function of system pressure

7.4.1 Effects of Deposition Pressure

In VFCVD it was found that when deposition pressure was changed from lower 10 Torr to higher 25 Torr it was increased in growth of diamond. It could be explained by accelerated increased number of reactive species such as atomic Carbon and hydrogen in the activated reaction chamber. It is also indicated that the SEM micrograph of the diamond film deposited on WC-Co substrate at 20 Torr and 25 Torr pressures region has produced higher density of diamond and larger grain size than lower at 10 Torr.

Figure 7.10 shows that deposition pressure Torr against growth of diamond grain size in μm. It is could be due to the generation of more carbon-containing radicals compared to atomic hydrogen. Figure 7.12a indicate that higher pressure reaching to 20 Torr will give good uniformity of larger grain size (~4 μm) of diamond and lower pressure 10 Torr will result in smaller grain size of diamond (<2 μm) (Fig. 7.11).

7.4.2 Effects of Methane Concentration

The effects of methane concentration on diamond growth were investigated. The diamond deposition is carried out on WC-Co dental bur substrate. Methane concentration was varied from 1, 1.5 to 2 % whilst other process parameters were kept constant.

Figure 7.13 shows the Raman spectra of three diamond films deposited at methane concentration of (a) 1 %, (b) 1.5 % and (c) 2 %. Under these conditions the

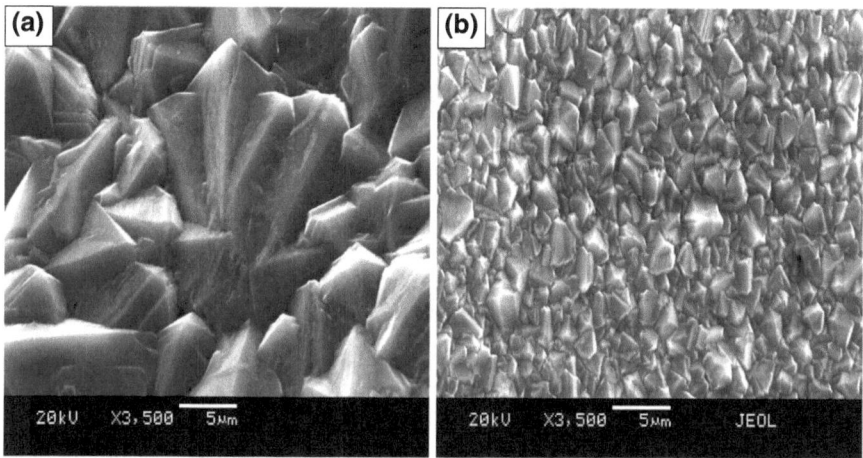

Fig. 7.12 **a** Larger grain size diamond with 20 Torr, **b** smaller grain size diamond with 10 Torr

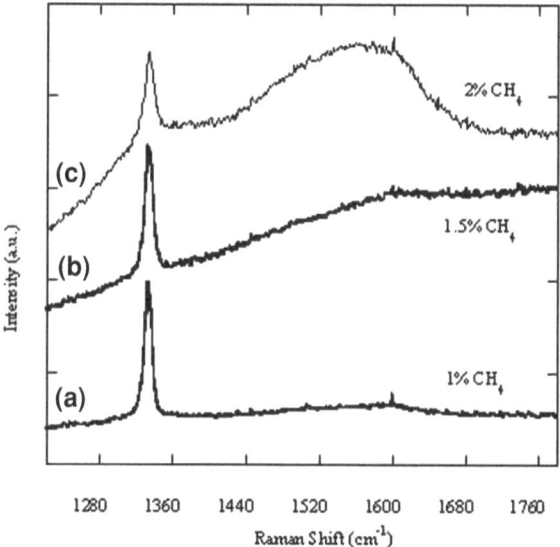

Fig. 7.13 Raman spectra of three different diamond films at **a** 1 %, **b** 1.5 % and **c** 2 % Methane CH_4

quality of diamond films decreased with higher methane concentration as indicated by the diamond 1,332 cm^{-1} peak intensity. Figure 7.14a–c shows that SEM micrographs corresponding to the as deposited diamond films with respective methane concentration.

It was evident that as methane concentration was increased, the carbon phase purity of VFCVD diamond films deteriorated and diamond grain density increased

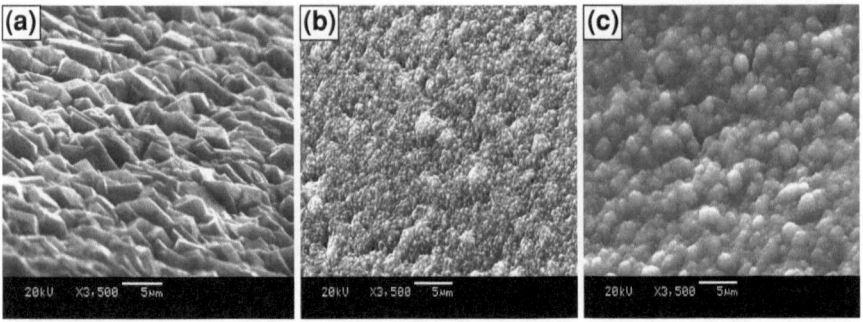

Fig. 7.14 SEM micrograph corresponding to as deposited CVD diamond films **a**, **b** and **c** respect to $CH_4\%$

and films looked smoother. This is as expected since more carbon-containing radical species would be generated. It also suggested that the etching effect of atomic hydrogen might become relatively insufficient, thus favouring the formation of growth defects. Furthermore, non-carbon phases are a consequence of higher methane concentration, as can be seen from broad absorption around $1{,}550~\text{cm}^{-1}$ in the Raman spectra.

7.5 Conclusions

The use of negative substrate biasing and chemical etching enhanced the nucleation density of diamond. Short bias times of the order of 10, 20 and 30 min were sufficient for the subsequent growth of quality diamond films. Smoother diamond films were obtained after biasing at higher voltages (-300 V). This may be due to secondary nucleation mechanisms of diamond on the deposited surfaces, which produced thicker films with an increased nucleation density. The substrate temperature affects the morphology of the as-grown diamond with larger crystal sizes growing at higher substrate temperatures.

References

Ager I, JW, Drory MD (1993) Quantitative measurement of residual biaxial stress by Raman spectroscopy in diamond grown on a Ti alloy by chemical vapor deposition. Phys Rev B 48(4):2601

Amirhaghi S et al. (2001) Diamond coatings on tungsten carbide and their erosive wear properties. Surf Coat Technol 135(2):126–138

Butler JE et al. (2009) Recent progress in the understanding of CVD growth of diamond. CVD Diamonds Electron Devices Sens:103–124

Cabral G et al. (2006) Diamond chemical vapour deposition on seeded cemented tungsten carbide substrates. Thin Solid Films 515(1):158–163

Cock AM et al. (1994) Comparison of two models of thin diamond film microhardness data to predict the hardness of CVD diamond. Diam Relat Mater 3(4–6):783–786

Deuerler F et al. (1996) Pretreatment of substrate surface for improved adhesion of diamond films on hard metal cutting tools. Diam Relat Mater 5(12):1478–1489

Filik J et al. (2006) Raman spectroscopy of diamondoids. Spectrochim Acta Part A Mol Biomol Spectrosc 64(3):681–692

Mankelevich YA, May PW (2008) New insights into the mechanism of CVD diamond growth: single crystal diamond in MW PECVD reactors. Diam Relat Mater 17(7):1021–1028

Matsumoto S et al. (1982) Vapor deposition of diamond particles from methane. Jpn J Appl Phys 21(4A):L183

May PW et al. (2010) Simulations of polycrystalline CVD diamond film growth using a simplified Monte Carlo model. Diam Relat Mater 19(5–6):389–396

Murakawa M, Takeuchi S (1991) Mechanical applications of thin and thick diamond films. Surf Coat Technol 49(1):359–364

Polini R et al. (2004) Cutting force and wear evaluation in peripheral milling by CVD diamond dental tools. Thin Solid Films 469:161–166

Polini R et al. (2010) Wear resistance of nano-and micro-crystalline diamond coatings onto WC-Co with Cr/CrN interlayers. Thin Solid Films 519(5):1629–1635

Schneider A et al. (2010) Enhanced tribological performances of nanocrystalline diamond film. Int J Refract Metal Hard Mater 28(1):40–50

Sein H et al. (2004a) Enhancing nucleation density and adhesion of polycrystalline diamond films deposited by HFCVD using surface treatments on co cemented tungsten carbide. Diam Relat Mater 13(4):610–615

Sein H et al. (2004b) Performance and characterisation of CVD diamond coated, sintered diamond and WC-Co cutting tools for dental and micromachining applications. Thin Solid Films 447–448:455–461

Wang BB et al. (2001) Theoretical analysis of ion bombardment roles in the bias-enhanced nucleation process of CVD diamond. Diam Relat Mater 10(9):1622–1626

Wang WL et al. (1999) Mechanism of diamond nucleation enhancement by electron emission via hot filament chemical vapor deposition. Diam Relat Mater 8(2):123–126

Xianglin LI (2008) High-rate diamond deposition by microwave plasma CVD, ProQuest

Chapter 8
VFCVD Diamond Dental Burs for Improved Performance

Abstract In this chapter the performance and life of dental burs coated with diamond has been investigated. The performance of various burs both coated with diamond films using VFCVD and un-treated burs have been compared. Results show that the diamond coated burs using VFCVD performed better in terms of life.

Keywords VFCVD · Chemical vapour deposition · Diamond coating · Life · Performance

8.1 Introduction

CVD diamond films have attracted considerable interest for cutting tool applications, including rotary tools and inserts due to their excellent physical and chemical properties (Ahmed et al. 2000; Afzal et al. 1998; Ali et al. 1999; Jones et al. 2003). However, deposition of adherent high quality diamond films to substrates such as cemented carbides, stainless steel and various metal alloys containing transition elements has proven to be a problem. In general, the adhesion of the diamond films to the substrates is poor and the nucleation density is very low (Leyendecker et al. 1991; Murakawa and Takeuchi 1991; Jackson et al. 2004). The influence of different metallic substrates on the diamond deposition process has been examined (Deuerler et al. 1996; Kupp et al. 1994; Bauer et al. 2003; Lee et al. 1998). The physical and chemical nature of the substrate has a crucial impact on diamond nucleation and its subsequent growth as described in Chap. 6.

The present work is concerned with diamond deposition onto tungsten carbide cemented with 6 wt% cobalt content. WC-Co substrates are suitable for diamond deposition, but their adhesion strengths to diamond films are relatively poor (Pines and Schulman 1979). The poor adhesion is related to the cobalt binder that is present to increase the toughness of the tool. Much effort has been directed at increasing the adhesion strength of diamond films to WC-Co substrates, including decarburizing the surface prior to deposition (Sein et al. 2004), seeding WC-Co with diamond powder and annealing prior to deposition (Polini et al. 2003), removing cobalt atoms at the surface using cobalt etching agents (Polo et al. 1997; Endler et al. 1999), and depositing an interlayer such as TiN as a diffusion barrier.

If these deficiencies can be overcome then CVD diamond coatings have the potential to considerably prolong the lifetime of WC-Co dental burs when applied to the machining of highly abrasive non-ferrous metallic alloys, borosilicate glass, porcelain or acrylic teeth, natural human teeth, and ceramic materials. Therefore, it is essential to pre-treat substrates for both significantly reducing the surface Co concentration and achieving a proper interface roughness, which represents a important step prior to the coating process (Kamiya et al. 2000). It was shown in Chap. 5 that chemical treatment using Murakami reagent and acid etching has been used successfully for removal of the Co binder from the substrate surface resulting in growth of adherent diamond films.

In this chapter, results of the investigation of diamond film deposited on etched and un-etched WC-Co dental burs and subsequent machining results on human teeth, borosilicate glass, and acrylic teeth are presented. Up to now no known research has been carried out on the characterisation and performance of VFCVD diamond coated dental burs.

8.2 Experimental

8.2.1 Tool Preparation

Two sets of laboratory tungsten carbide (WC-Co) dental burs (AT23 LR) with fine WC grain sizes (1 μm) 20–30 mm in length and 1.0–1.5 mm in diameter [supplied by Metrodent Ltd, UK.], were used as substrates. The two-step Murakami/acid etching procedure outlined in Sect. 5.3.1 was used on all substrates. By using SEM and EDX characterized the etched surfaces of substrates.

8.2.2 CVD Diamond Deposition onto Dental Burs

Diamond films were deposited onto the cutting edge of the burs. The coiled tantalum filament (0.5 mm diameter) was held vertically within the vacuum deposition chamber and the dental burs were positioned centrally and coaxially within the coils of the filament such that the cutting edges were 5 mm from the filament coils. Prior to deposition, the tantalum filament was carburised for 30 min with 3 % CH_4 with excess hydrogen. Standard deposition conditions described

Fig. 8.1 Dental bur drilling machine

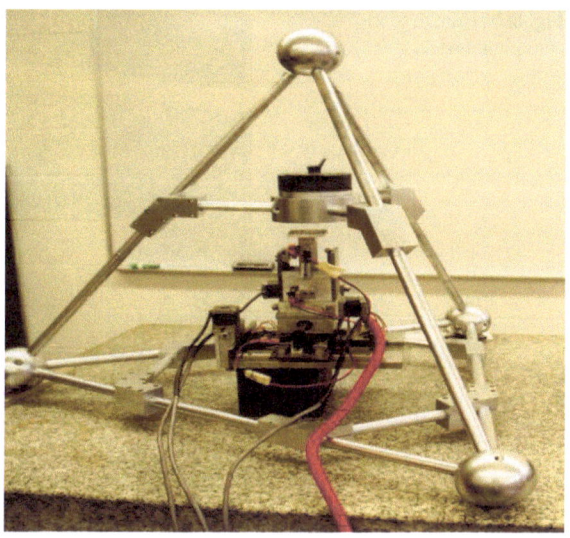

in Sect. 3.3 were used. The deposition time and pressure in the vacuum chamber were 5–15 h and 20 Torr, respectively.

8.2.3 Dental Bur Machining: Drilling Experiments

In order to examine the cutting performance of the diamond coated dental burs a variety of clinical and laboratory materials such as borosilicate glass, acrylic teeth, and natural human teeth were drilled. The drilling unit (Fig. 7.1) was constructed and used by collaborators at Purdue University, USA. It was constructed with a water-cooling system so that maximum spindle speeds of 250,000 revolutions per minute (rpm) feed rates of between 5–20 μm per revolution, and cutting speeds in the range 100–200 m/min for drilling with dental burs could be achieved.

After the dental burs were coated and examined for adhesion, the coated burs were used to machine a number of dental materials. The coated burs were compared with uncoated burs to distinguish them in terms of their drilling behaviour. The machine shown in Fig. 8.1 was constructed using three principal axes each controlled using a dc motor connected to a Motionmaster^{TM} controller. A laser light source was focused on the rotating spindle in order to measure the speed of the dental bur during drilling.

The flank wear of the burs were analysed by SEM at selected machining time intervals of between 1 and 3 min. Prior to SEM analysis the diamond coated burs were ultrasonically washed with 6M H_2SO_4 solutions to remove any unwanted machining material, which eroded the surface of the VFCVD diamond coated bur. For comparison, conventional diamond sintered burs with different geometry were also tested on the same substrate materials.

Fig. 8.2 Human tooth
clamping device

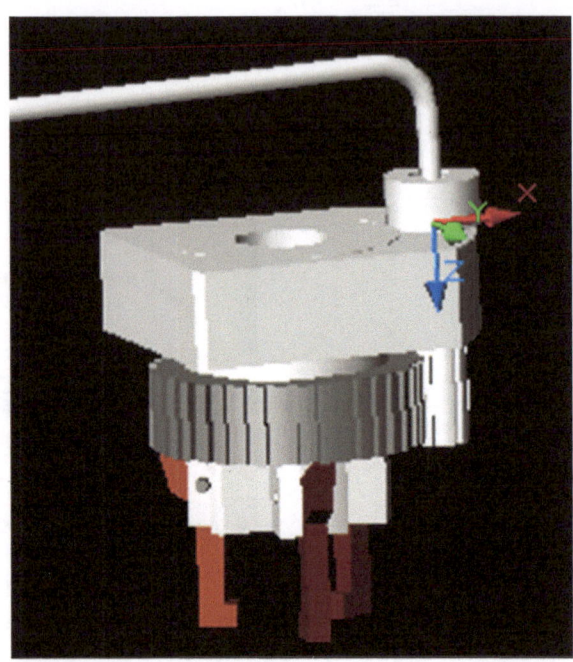

Fig. 8.3 Tooth clamping
device and driving stage used
to locate clamping device to
the surface of the tooth

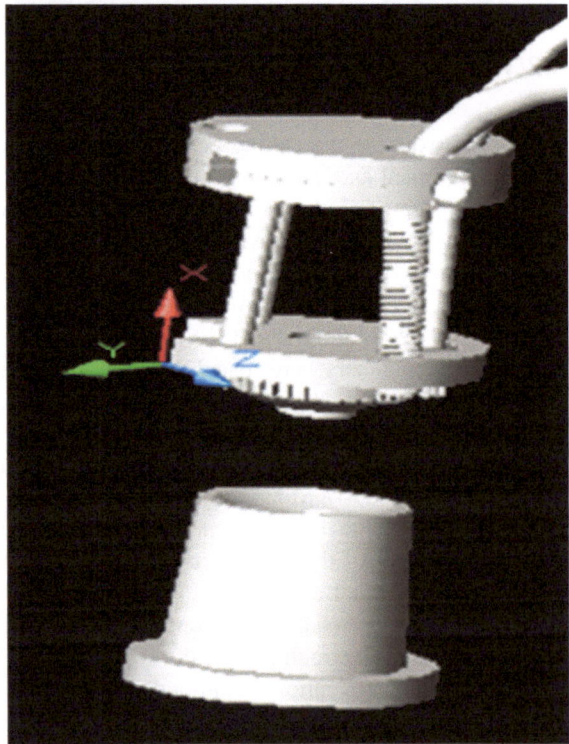

Fig. 8.4 Air operated
spindle unit attached to the
clamping device and driving
unit attached firmly to the
tooth

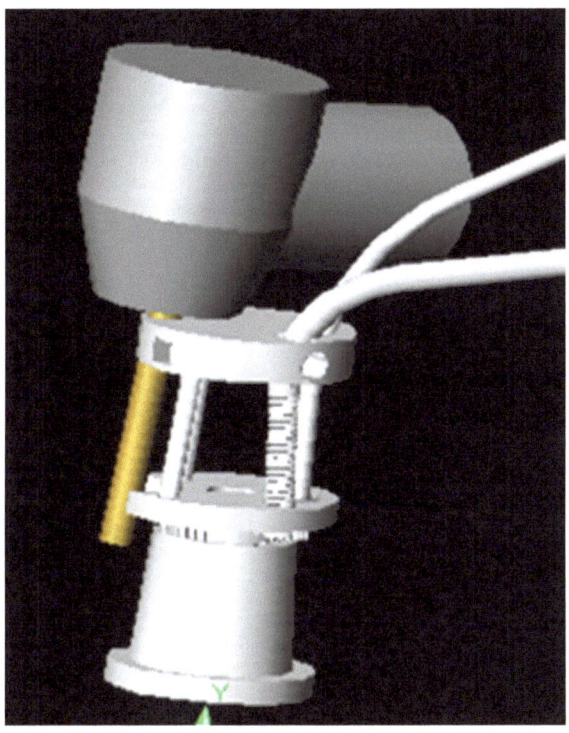

8.2.4 Dental Bur Machining: Machining Experiments on Human Teeth

To examine the machining characteristics of coated and uncoated dental burs, a specially constructed clamp was developed to locate over the tooth to prepare it for the location of a crown. Figures 8.2, 8.3 and 8.4 shows the construction of the clamping device.

The clamping device is located onto the tooth to be machined and allows the tooth to be machined by incorporating a wire driven driving mechanism that attaches itself onto the clamp so that the dental bur can rotate at the appropriate cutting speed. Figure 8.3 shows the driving mechanism attached to the clamping device and a tooth. The driving mechanism is attached to the air operated hand piece that provides the power to drive the mechanism, clamp, and dental bur. Figure 8.4 shows the construction of the full assembly that allows machining of the tooth-materials to take place. The bur was rotated at 20,000–30,000 rpm, with a feed rate of between 0.2–0.5 mm/revolution without water-cooling. The uncoated and coated dental burs were compared with sintered diamond burs machining acrylic teeth, and borosilicate glass used to simulate the machining of teeth. The flank wear of the dental burs was estimated from SEM.

Fig. 8.5 Human tooth-
clamping devices

Fig. 8.6 Tooth clamping
device and driving stage used
to locate clamping device
onto surface of the tooth

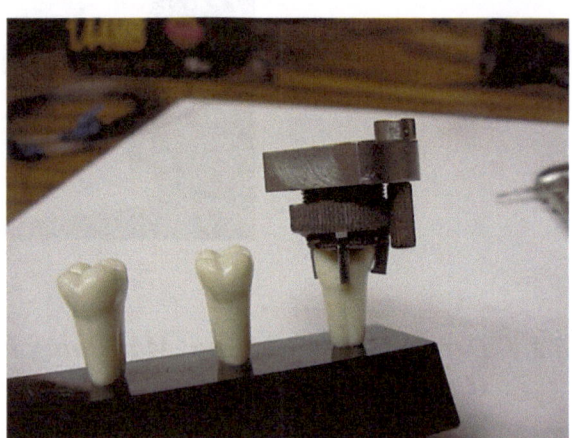

8.3 Dental Bur Machining: Machining Experiments

To examine the machining characteristics of coated and uncoated dental burs, a specially constructed clamp was developed to locate over the tooth to prepare it for the
location of a crown. Figure 8.5 shows the basic construction of the clamping device.

The clamping device is located onto the tooth to be machined and allowed
the tooth to be machined by incorporating a wire driven driving mechanism that
attaches itself onto the clamp so that the dental bur can rotate at the appropriate
cutting speed. Figure 8.5 shows the driving mechanism attached to the clamping
device and a tooth. The driving mechanism is attached to the air operated hand
piece that provides the power to drive the mechanism, clamp, and dental bur.
Figure 8.6 shows the construction of the full assembly that allows machining of
the tooth to take place. The bur was rotated at 20,000–30,000 rpm, with a feed rate

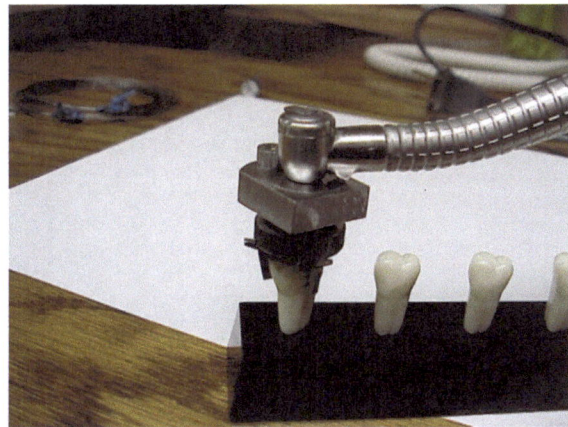

Fig. 8.7 Air operated spindle unit attached to the clamping device and driving unit attached firmly to the tooth

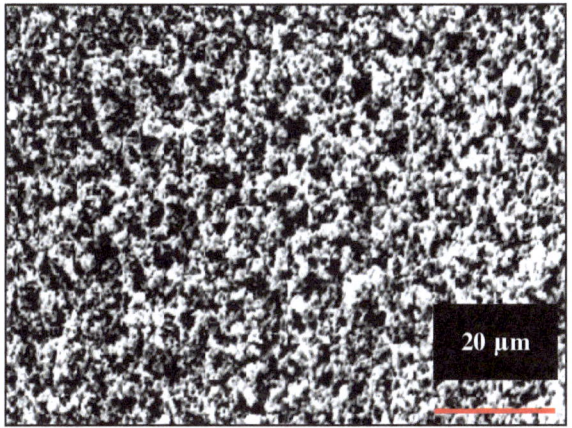

Fig. 8.8 Cutting edge of a WC-Co dental bur after etching with acid and Murakami's solution

of between 0.2–0.5 mm/revolution without water cooling (Fig. 8.7). The uncoated and coated dental burs were compared with sintered diamond burs for machining acrylic teeth, human teeth, and borosilicate glass. Flank wear of the dental burs was estimated using a scanning electron microscope. Again, H_2SO_4 solution was used to remove detritus from the surfaces of the dental burs.

8.4 Results and Discussion

8.4.1 Effects of Substrate Preparation on Diamond Deposition

The effects of the Murakami/acid chemical etching can clearly be seen on the WC-Co substrate surfaces. Figure 8.8 shows that cobalt etching was successfully

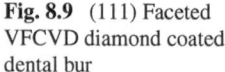 **Fig. 8.9** (111) Faceted VFCVD diamond coated dental bur

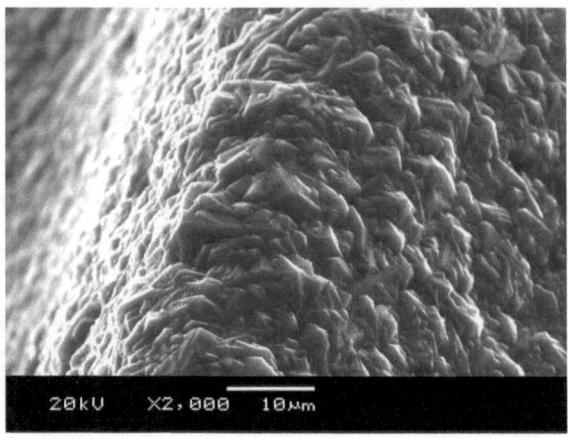

achieved because no cobalt peaks could be detected in the EDX spectrum. In addition the etching process produced a roughened surface.

Diamond layers were deposited on pre-treated WC-Co dental burs. Adherent diamonds on WC-Co dental burs consisting of mainly (111) faceted diamond crystals were deposited (Fig. 8.9). Micrographs of the burs show that the diamond film is uniformly coated onto the complex 3-D substrate surface. The modified filament arrangement produced a uniform and dense diamond coating even though the substrate is non-planar with a complex geometry. The morphology of the diamond surface of the dental bur is rough making the bur extremely desirable for dental machining applications.

The coated dental bur was cut in order to study the cross section of the tool as can be seen in Fig. 8.10. It was found that the coating is thicker at the cutting teeth with average thickness of about 15–17 μm (growth rate of 1.1 μm/h) due to the slightly higher temperature at the bur tip because cutting teeth is closer to the filament coil. At the base of the bur the heat is carried away faster and therefore it is at a lower temperature giving rise to lower growth rates and hence thinner films, at about 8 μm in thickness. The thicker coating at the tip is expected to give the tool longer life. Further work is required to study the effects of film thickness at the tool tip and at the base on tool performance and lifetime.

8.5 Performance Experiments

Figure 8.11 shows a SEM of a conventional polycrystalline diamond (PCD) sintered bur. The diamond particles are imbedded to surface with a suitable binder matrix material, in this case nickel.

Typically the surface is inhomogeneous and sizes of particles are range from 50 to 200 μm causing considerable variation in the cutting performance of the tool. An important factor that could affect the final performance of the dental bur

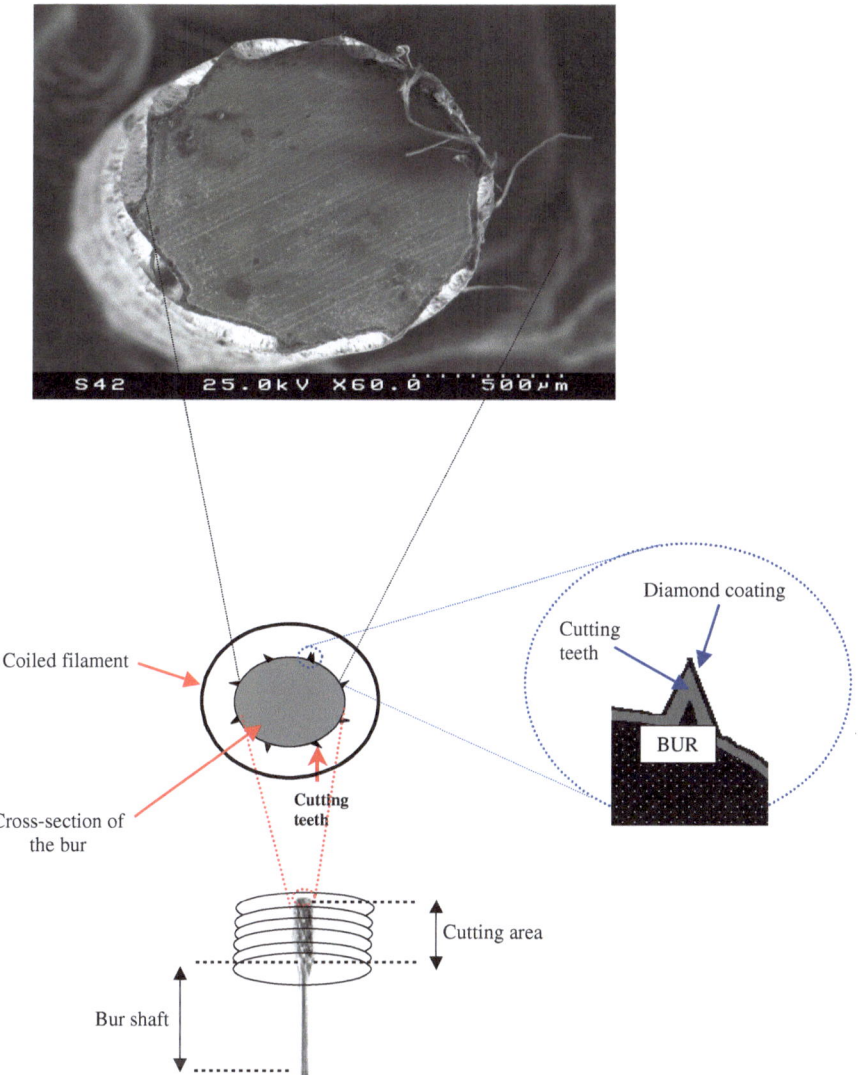

Fig. 8.10   Cross section of dental burs coated with VFCVD diamond

is the adhesive toughness of the diamond on the substrates. Endler et al. (1996, 1999) and Kamiya et al. (2000) have developed a new method for the quantitative evaluation of the adhesive toughness of diamond films onto Co-cemented WC substrates. They found that the adhesive toughness of diamond on WC to be in the range of 20–37 J/m^2. Commercial burs exhibited much higher adhesive toughness than flat substrates due to the large surface roughness and the absence of interfacial voids. This factor needs to be investigated in detail for non-planar dental burs.

Fig. 8.11 Inhomogeneous
surface of PCD diamond
sintered bur

The effectiveness of using VFCVD coated dental burs was measured by comparing uncoated burs, VFCVD dental burs, and polycrystalline diamond (PCD)
sintered diamond burs when drilling and machining extracted human teeth, acrylic
teeth, and borosilicate glass.

8.5.1 Experimental

A sequence of fifty drillings was employed in each drilling experiment. The sharp-
ness and initial condition of the burs were inspected visually after the burs had
drilled ten holes in sequence. An abrading coefficient of drilling, C_a, has been
defined as a quality criterion for small drilling tools (Sein et al. 2004). It is defined
as the ratio between the bur's total abraded area, S, and the effective coated area
of the bur used during the drilling process. The effective coated area is given by
the difference of the nominal coated bur area, D_b, and the area of the bur material
consumed or removed during drilling, W_b. C_a is defined by the following equation:

$$C_a = S / (D_b - W_b) \qquad (8.1)$$

A high quality coated dental bur is one that produces accurate drilling that has
an area S close to D_b and does not lose its coating during the machining process,
i.e. $W_b \approx 0$. The way W_b, C_a and D_b are defined is given in Sein H et al (2004).
The cutting will therefore have an abrading coefficient close to unity. One must
remember that the quality of machining is dependent on the cutting speed, V_c,
defined by the ratio of the drilling depth by the drilling duration. A comparative
factor of merit for the dental bur is defined as:

$$F = C_a / V_c \qquad (8.2)$$

where F, the figure of merit, is directly related to the lifetime of the dental bur for
a specific drilling process. A figure of merit (F) equal to unity represents an ideal
bur where there is zero wear occurring on the burs.

Fig. 8.12 Figure of merit as
a function of number of holes
drilled for dental burs drilling
borosilicate glass

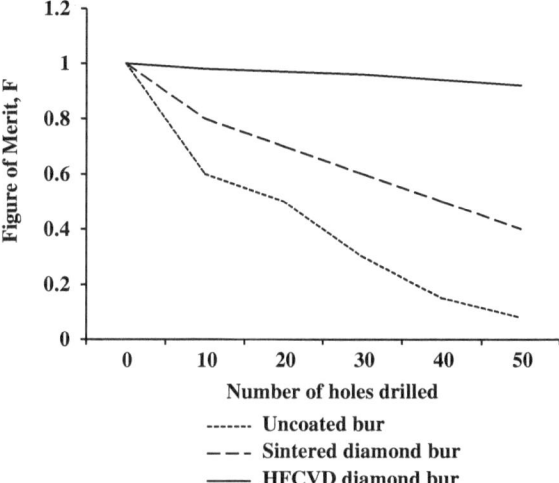

Fig. 8.13 Figure of merit as
a function of number of holes
drilled for dental burs drilling
acrylic material

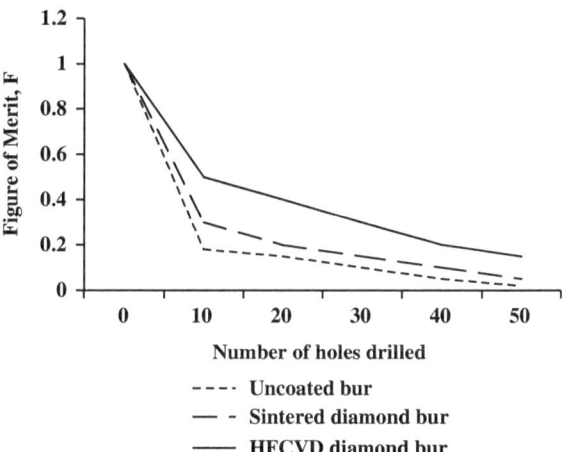

8.5.2 Results and Discussion

Figures 8.12, 8.13 and 8.14 show the results of drilling the dental materials with
the three types of burs on borosilicate glass, acrylic tooth materials and human
teeth, respectively.

For all the materials investigated including acrylic, borosilicate glass and
human tooth material it is evident that the best-performing bur in term of figure of
merit were the VFCVD diamond coated burs on the WC-Co substrates and TiN.

Figure 8.15 shows a micrograph of uncoated WC-Co dental bur tested on the
borosilicate glass using the same machining conditions. The uncoated WC-Co
burs displayed flank wear along the cutting edge of the bur. The areas of flank
wear were investigated at the cutting edge of the dental bur. Figure 8.12 shows

Fig. 8.14 Figure of merit as a function of number of holes drilled for dental burs drilling human tooth material

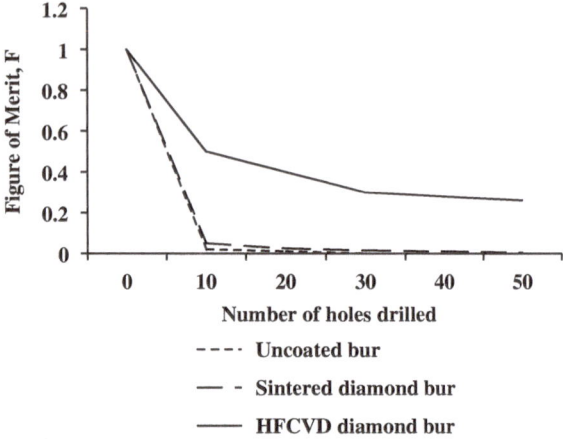

Fig. 8.15 Worn cutting edge uncoated WC-Co dental bur after drilling borosilicate glass

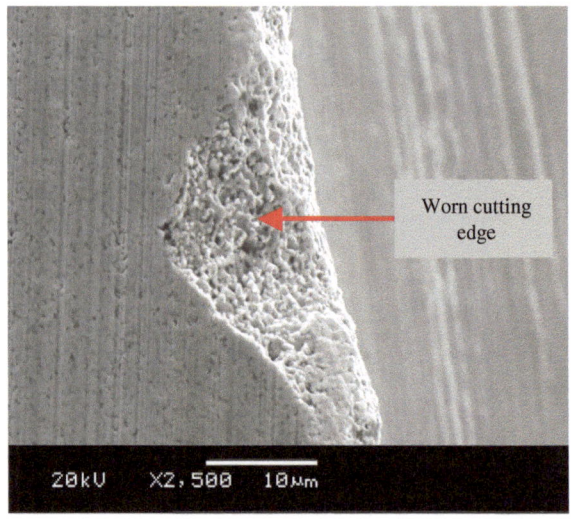

flank wear as a function of cutting time when drilling borosilicate glass. It is evident that the action of machining causes high rates of flank wear on the cutting edge of dental bur. Therefore, the cutting edges of WC-Co dental burs should have a minimum thickness of VFCVD diamond, which will enhance not only quality of cutting but also prolong the life of the bur.

Figure 8.16 shows the surface of a sintered diamond bur after being tested on borosilicate glass at a cutting speed of 30,000 rpm for 5 min with an interval at every 30 s. It is clear that there is significant removal of diamond particles from the surface of the tool after 50 holes. As expected there is deterioration of the abrasive performance of the PCD sintered diamond dental burs. Borges et al. also reported that the significant loss of diamond particles occurred during cutting with the commercial sintered diamond bur (Borges et al. 1996, 1999). In addition,

Fig. 8.16 PCD sintered bur
after drilling on borosilicate
glass

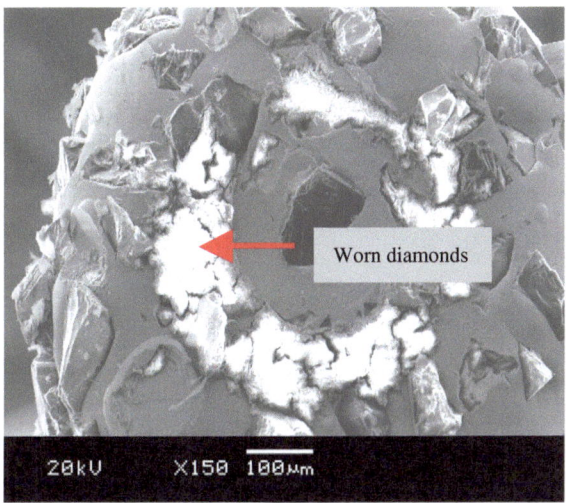

Fig. 8.17 Cutting edge of
dental bur after drilling on
borosilicate glass

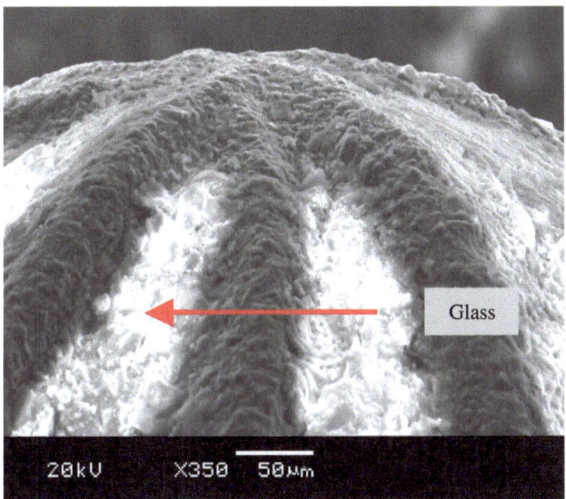

the nickel binder shows major defects generated by pulled-out particles (Pines and
Schulman 1979).

Figures 8.17 and 8.18 show an SEM images of a VFCVD diamond coated bur
after drilling experiments on borosilicate glass and acrylic teeth, respectively, for
5 min at a cutting speed of 30,000 rpm.

After machining, it is evident that the diamond films are still intact on the pre-
treated WC substrate and diamond coating displayed good adhesion. Also there is no
indication of diffusion wear after the initial test for 50 holes. However, the machined
materials such as glass pieces erode the cutting edge of the diamond dental bur as
adhesive wear was observed (Fig. 8.17). After conducting experiments on acrylic teeth

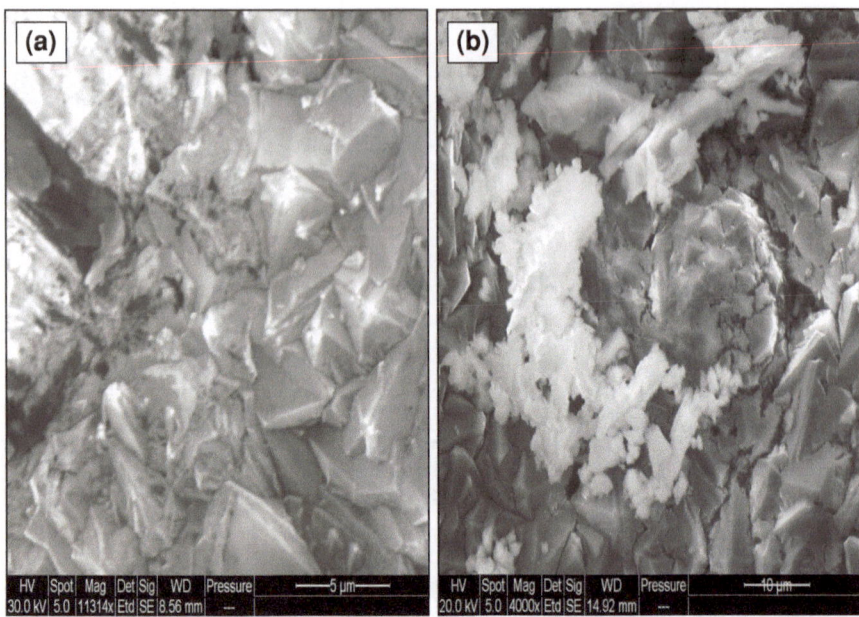

Fig. 8.18 Magnified image of the VFCVD diamond coated dental bur before (**a**) after (**b**) drilling acrylic tooth material

Fig. 8.19 SEM micrograph of diamond coated dental bur at a cutting speed of 250,000 rpm

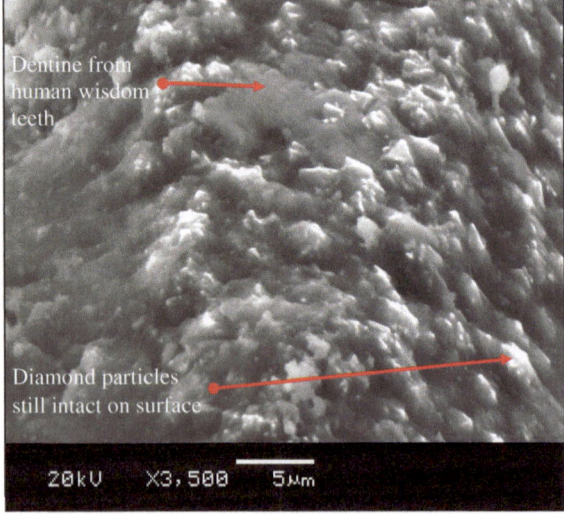

the mechanisms of wear involves adhesion as well as abrasion. Figure 8.18 shows that inorganic fillers from acrylic teeth adhered to the cutting tool surface in localised areas when increased rate of abrasion was used (Jackson et al. 2007; Shimosato 2003).

Natural human teeth were also drilled using a diamond coated dental bur. Previous studies have indicated that teeth should not be used for reduction tests

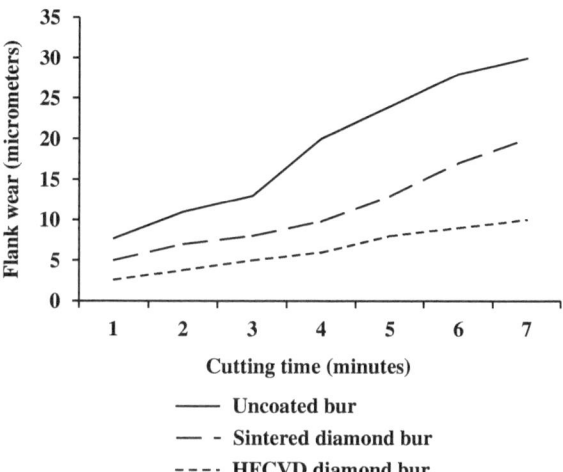

Fig. 8.20 Flank wear of burs machining borosilicate glass

because of the differences in hardness between enamel and dentine (Knoop hardness data: enamel, 250–500 kg mm^{-2}; dentine, 50–70 kg mm^{-2}). Cuts were made in the central groove of the teeth with the burs. This permitted cutting three grooves in each tooth. Figure 8.19 shows the SEM of a diamond coated WC-Co dental bur after testing on extracted teeth. It is evident from the micrograph that the tooth materials such as dentine clog up on the bur reducing its abrasive performance.

An area where this could be important is the surface treatment of small dental tools such as dental burs, which are used in the dental laboratory and clinical surgery for removing unwanted material from teeth. The negatively bias assisted and diamond coated WC-Co dental burs were tested with human teeth in order to observe their adhesive strength of diamond particles on the surface. The coated burs have been tested on extracted human wisdom teeth, which have a difference in hardness between enamel and dentine. The teeth were cut in a bench device using an ultra high-speed hand piece (air rotor). A frequency meter monitored the speed of the hand piece, which was between 200,000–250,000 rpm. Water-cooling device was used to prevent the tool overheating. The SEM micrograph of tested dental bur shows that the diamond film was still intact after the application of highly abrasive cutting at a speed of 250,000 rpm for 2 min (Fig. 8.19). These results are extremely encouraging and clearly demonstrate the extreme toughness and durability of the diamond films using our modified VFCVD system.

A series of machining experiments were also conducted using uncoated, VFCVD dental burs, and sintered diamond burs when machining extracted human teeth, acrylic teeth, and borosilicate glass. The teeth were machined using the apparatus shown in Figs. 8.6 and 8.7. The life of the burs in the machining sense was measured by comparing the amount of flank wear exhibited by each type of dental bur. The flank wear was measured at time intervals of 2, 3, 4, 5, 6, and 7 min machining duration using SEM micrographs (Harano et al. 2012). Figures 8.20, 8.21 and 8.22 show the flank wear measurements for each bur machining different dental materials.

Fig. 8.21 Flank wear of
burs machining acrylic tooth
material

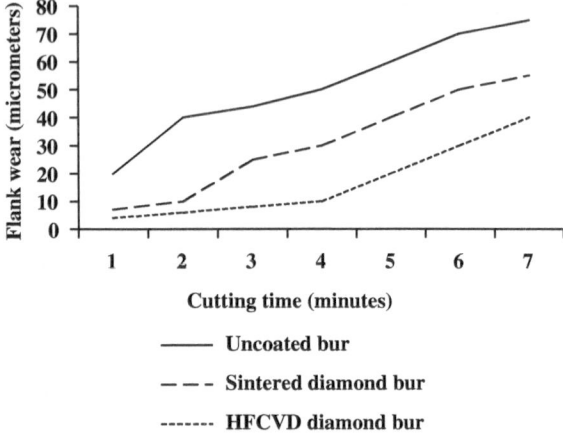

Fig. 8.22 Flank wear of
burs machining human tooth
material

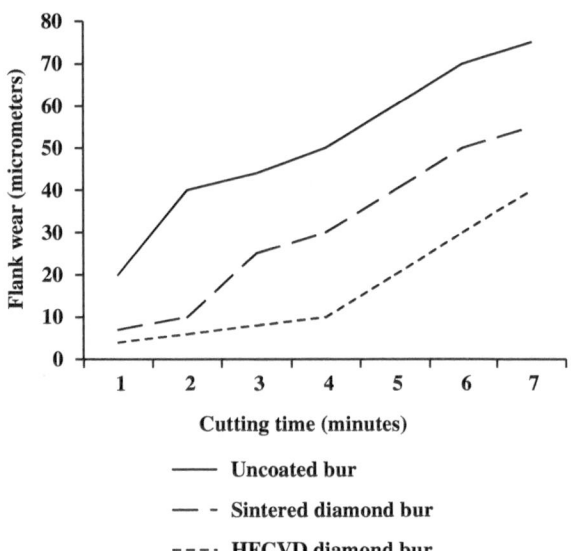

Again, the dental burs were examined using optical and scanning electron microscopic techniques and were found to observe similar trends as burs associated with drilling experiments close.

Three different dental bur (a) bias assisted diamond dental bur (b) non-bias assisted diamond dental bur (c) uncoated WC-Co dental bur are used for machining work piece e.g. cobalt-chrome alloy.

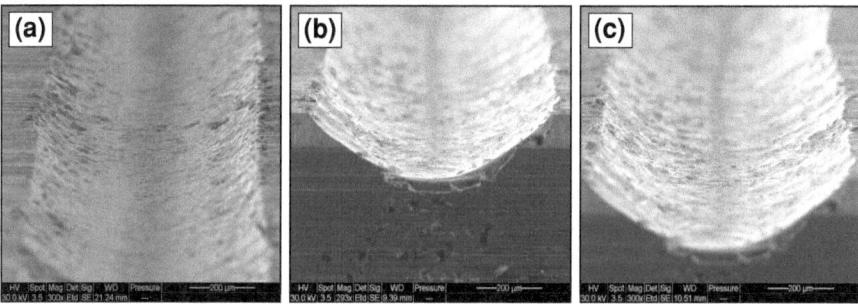

Fig. 8.23 Comparison of different dental bur performance. **a** Bias assisted diamond coated dental bur. **b** Non-bias diamond coated dental bur. **c** Uncoated WC-Co dental bur

After machining the material, the surface characteristics was investigated by SEM and observations indicated that bias-assisted diamond coated bur gave smoother surface due to smooth diamond film on bur. Other uncoated bur and non-bias assisted diamond bur have shown rough cut on surface (Fig. 8.23).

8.6   Conclusions

The etching treatment of the surface of the substrates described here to remove the cobalt surface layer results in good adhesion of the diamond film to the WC-Co substrate. The PCD sintered diamond bur loses significant proportions of embedded diamond particles during the abrasive machining procedure.

VFCVD diamond coated diamond burs remain intact with the potential of prolonging tool life. A thicker coating of CVD diamond at the cutting edges is expected to give longer bur life, and a much better quality of drilling and machining. The performance and lifetime of VFCVD coated dental bur is far superior to the sintered bur and the uncoated WC-Co bur. Further work is required to study the effects of diamond film adhesion and thickness of the coating at the cutting edge on dental tool life.

References

Afzal A et al (1998) HFCVD diamond grown with added nitrogen: film characterization and gas-phase composition studies. Diam Relat Mater 7(7):1033–1038

Ahmed W et al (2000) CVD diamond: controlling structure and morphology. Vacuum 56(3):153–158

Ali N et al (1999) Role of surface pre-treatment in the CVD of diamond films on copper. Thin Solid Films 355:162–166

Bauer CE et al (2003) A comparative machining study of diamond-coated tools made by plasma torch, microwave, and hot filament techniques. Sadhana 28(5):933–944

Borges CFM et al (1999) Dental diamond burs made with a new technology. J Prosthet Dent 82(1):73–79

Borges CFM et al (1996) Adhesion of CVD diamond films on silicon substrates of different crystallographic orientations. Diam Relat Mater 5(12):1402–1406

Deuerler F et al (1996) Pretreatment of substrate surface for improved adhesion of diamond films on hard metal cutting tools. Diam Relat Mater 5(12):1478–1489

Endler I et al (1999) Preparation and wear behaviour of woodworking tools coated with superhard layers. Diam Relat Mater 8(2):834–839

Endler I et al (1996) Interlayers for diamond deposition on tool materials. Diam Relat Mater 5(3–5):299–303

Harano K et al (2012) Cutting performance of nano-polycrystalline diamond. Diam Relat Mater 24:78–82

Jackson MJ et al (2004) Diamond-coated cutting tools for biomedical applications. J Mater Eng Perform 13(4):421–430

Jackson MJ et al (2007) Dental drilling in severe and demanding environments. Int J Nano Biomater 1(2):165–183

Jones AN et al (2003) The impact of inert gases on the structure, properties and growth of nanocrystalline diamond. J Phys Condens Matter 15(39):S2969

Kamiya S et al (2000) Quantitative determination of the adhesive fracture toughness of CVD diamond to WC–Co cemented carbide. Diam Relat Mater 9(2):191–194

Kupp ER et al (1994) Interlayers for diamond-coated cutting tools. Surf Coat Technol 68:378–383

Lee D-G et al (1998) Novel method for adherent diamond coatings on cemented carbide substrates. Surf Coat Technol 100:187–191

Leyendecker T et al (1991) Industrial application of crystalline diamond-coated tools. Surf Coat Technol 48(3):253–260

Murakawa M, Takeuchi S (1991) Mechanical applications of thin and thick diamond films. Surf Coat Technol 49(1):359–364

Pines MS, Schulman A (1979) Characterization of wear of tungsten carbide burs. J Am Dent Assoc 99(5):831–833

Polini R et al (2003) Dry turning of alumina/aluminum composites with CVD diamond coated Co-cemented tungsten carbide tools. Surf Coat Technol 166(2):127–134

Polo MC et al (1997) Nucleation and initial growth of bias-assisted HFCVD diamond on boron nitride films. Diam Relat Mater 6(5):579–583

Sein H et al (2004) Performance and characterisation of CVD diamond coated, sintered diamond and WC–Co cutting tools for dental and micromachining applications. Thin Solid Films 447:455–461

Shimosato Y (2003) Low pressure vacuum carburizing and accelerated gas carburizing. In: Heat treating and surface engineering, proceedings of the 22nd heat treating society conference and the 2nd international surface engineering congress